Lecture Notes in Physics

For information about Vols. 1–131, please contact your bookseller or Springer-Verlag.

Lecture Notes in Physics

Edited by H. Araki, Kyoto, J. Ehlers, München, K. Hepp, Zürich
R. Kippenhahn, München, H. A. Weidenmüller, Heidelberg
and J. Zittartz, Köln

204

Yoshio Waseda

Novel Application of Anomalous (Resonance) X-ray Scattering for Structural Characterization of Disordered Materials

Springer-Verlag
Berlin Heidelberg GmbH 1984

Author

Yoshio Waseda
Research Institute of Mineral Dressing and Metallurgy (SENKEN)
Tohoku University, Sendai 980, Japan

ISBN 978-3-540-13359-9   ISBN 978-3-540-38910-1 (eBook)
DOI 10.1007/978-3-540-38910-1

Originally published by Springer-Verlag Berlin Heidelberg New York Tokyo in 1984

2153/3140-543210

PREFACE

The past ten years or so have seen a remarkable growth in the subject of the structure and properties of disordered materials at a microscopic level. This is because of the novelty of the physics, mainly relating to the particular **non-periodicity** in their atomic arrangements, and also because of the technological potential of some of these disordered materials for almost immediate use as soft magnetic elements and electronic devices. However, our present understanding of the physical or chemical phenomena in disordered materials is often far from complete, for several reasons. One of the main reasons is that in multi-component disordered materials it is relatively difficult to determine the fine structure, say, the compositional short range order (CSRO), which must be described by the so-called partial radial distribution functions (RDFs).

Several techniques have been applied to determine the CSRO or partial RDFs in multi-component disordered materials, but the accuracy of these results does not appear to be sufficient to allow definite interpretation of the data in some cases. The use of **anomalous (resonance) x-ray scattering** will, in the author's view, bring about a significant breakthrough in such difficulties by permitting the accurate evaluation of the CSRO in multi-component disordered materials, particularly when coupled with a high intensity white x-ray source such as the synchrotron radiation . However, a number of developments are still required before it can be accepted as a very reliable technique for structural characterization of disordered materials.

Reflecting these focal problems, the aim of this monograph is to provide an extended introductory treatise on the novel application of anomalous (resonance) x-ray scattering to structural characterization of disordered materials. The subjects matter is treated selectively rather than comprehensively, and its level is commensurate with that of a graduate course in physics and materials science. It is also directed towards persons in other area, such as chemistry and ceramics, who wish to become acquainted with the relatively new subjects of disordered materials. The text gives a critical, up-to-date evaluation covering present problems and future directions in novel applications of the anomalous x-ray scattering, primarily for structural studies of disordered materials. It also gives relevant fundamental information, such as the energy dependence of the anomalous dispersion factors for various elements, compiled in the appendices. This information is always required, **not only for disordered materials, but also for crystalline materials,** to carry out successful anomalous x-ray scattering, but it is not covered in previously published specialized monographs and common databooks for x-ray crystallography. The author, therefore, believes that this monograph, with many references, provides an adequate help and guide for both specialists and non-specialists.

Many people have helped, directly and indirectly, in writing this text. The author is deeply indebted to Professors S.Tamaki, S.Hosoya, H.Iwasaki, and T.Egami for valuable discussion regarding various subjects of disordered materials. He is also grateful to Professors A.Yazawa, Y.Shiraishi and M.Ohtani for their encouragements in the series of research projects on the structure of disordered materials at the Research Institute of Mineral Dressing and Metallurgy (SENKEN), Tohoku University.

A significant part of Section 8.3 is based on the recent results of a collaboration involving the author and S.Aur, D.Kofalt, T.Egami, H.S.Chen, B.K.Teo and R.Wang. Their contributions as well as the dedicated service and advice given by the staff at Cornell High Energy Synchrotron Source (CHESS), Cornell University, particularly Professor B.W.Batterman, Drs. D.Bilderback and D.Mills, are gratefully acknowledged. A part of this book was written during the 1982-1983 academic session while the author was staying at the Department of Materials Science and Engineering, University of Pennsylvania. He wishes to acknowledge the members of the Department for their hospitality at that time, and also to thank the National Science Foundation for partial support.

The author is also grateful to Professors K.T.Aust, H.Araki, and Mr. D.Kofalt for reading the manuscript and making valuable suggestions. Many thanks are due to Dr. H.Ohta and Mr. T.Saito who prepared figures of the energy dependence of the anomalous dispersion factors, for 96 elements compiled in the Appendix. The author is also indebted to many sources for the materials in this article. These have been cited in the text with reference. Finally, the patience of his wife and children, tasked during the long time needed to concentrate on writing of this monograph, is very gratefully recognized, although the dedication of this work to them is indeed small compensation for their many sacrifices.

CONTENTS

CHAPTER 1

A BRIEF BACKGROUND OF THE PRESENT REQUIREMENT FOR STRUCTURAL CHARACTERIZATION OF DISORDERED MATERIALS

The physics and chemistry of **DISORDERED MATERIALS,** in which the atomic arrangement is not spatially periodic in contrast to the case of crystalline materials, is now well-recognized as an important and promising branch for research and development. The typical examples of disordered materials are liquids and amorphous solids of condensed matter. Current interest in this relatively new and rapidly growing field mainly arises from the development of a new class of disordered materials known as metallic glasses and amorphous semiconductors, because of their technological potentials of some of these materials for application such as soft magnetic elements and electronic devices. Thus, many new advances have been made only recently, although this research field itself has long been studied in the past.

On the other hand, liquid metals and slags which are mainly oxide mixtures are known to play a significant role in many metallurgical processes (see for example, Richardson 1974). In addition, the calcium ferrite-base slags, without silica, have recently received much attention, because their physical and chemical properties indicate promise in reducing several difficulties in non-ferrous metallurgical operations with the usual silicate-base slags (see for example, Yazawa et al. 1981). Some liquid alkali metals are potential heat-transfer medium in nuclear reactor process. Their growing technological importance and the novelty of physics mainly related to the non-periodicity in the atomic arrangement of disordered materials (Ziman 1979) have led to an increasing need for the better describing their atomic scale structures and the better understanding of their various properties at a microscopic level.

The description of the atomic scale structure of disordered materials usually employs the **radial distribution function (RDF)** indicating the probability of finding another atom from an origin atom as a function of radial distance obtained by sphereical averaging (see for example, Hill 1956). The information given by the **RDF** is only one-dimensional but it does give almost unique quantitative information describing the atomic arrangements in disordered materials. The techniques of x-ray, neutron and electron diffraction have long been used to characterize the atomic scale structure of a variety of materials, particularly x-ray diffraction is well-known as one of the reliable tools for determining the RDF of disordered materials. However, the structural studies for disordered materials except for one-component systems are far from complete and they have had still relatively little impact on a direct link between the atomic scale structure and their characteristic properties, as already established in the case of crystalline materials, for several reasons. One of the most important reasons is that it is relatively difficult to determine the fine structure, say the near-neighbor atomic correlations of the individual chemical constitutents, frequently referred to as the **compositional short range order (CSRO)** well described by the **partial RDFs** , in multi-component disordered materials and thus it is hard to construct a realistic three-dimensional model structure in terms of the RDF data alone.

The supreme importance of a knowledge of fine structure has frequently been emphasized using various experimental results. For example, small changes in structure of disordered metallic materials (Chen 1980, Egami 1981a) appear to have significant consequence on the electronic and magnetic properties, although their basic structural features are well-characterized by the dense random packing of con-stituent metallic elements (Bernal 1959, Cargill III 1975).

Several techniques for extracting the Partial RDFs were proposed some time ago (see for example, Keating 1963) and a large amount of experimental and theoretical effort have been devoted to this particu-lar research field. For example, neutron diffraction using various isotopes, a combination of x-ray and neutron diffractions, the ex-tended x-ray absorption fine structure (EXAFS) measurement and multi-wavelength diffraction making use of **anomalous (resonance) x-ray scat-tering** have been applied to determine the CSRO of multi-component disordered materials in terms of the partial RDFs (see for example, Wagner 1978, Waseda 1980). More detailed discussion for a variety of these possible techniques and their relative merits and demerits will be given in the later section. However, the accuracy of these results does appear often not to be sufficient enough to allow quantitative discussion in detail, even for a binary disordered system and thus the available information on the CSRO is still limited for only a small number of compositions. Nevertheless, **the anomalous (resonance) x-ray scattering** technique by applying the so-called anomalous dispersion effect near the absorption edge of the constituent elements (see for example, Krogh-Moe 1966, Ramesh and Ramaseshan 1971) have recently received much attention for several reasons. The major reason is that

3

the use of the intense white radiation source from a synchrotron radiation coupled with the energy sensitive solid state detector can markedly improve the accuracy of this technique (Bienenstock 1975, Egami et al. 1978, Waseda 1981) and the synchrotron radiation sources produced by multi-GeV electron storage rings are now available in USA, Germany, England, France, Italy, Japan etc.

The anomalous x-ray scattering also enables us to offer information about local chemical environment of the desired elements, which is of course quite important for quantitative discussion of particular properties of disordered materials at a microscopic level. Such environmental structural information obtained by this technique is very similar to the results by the EXAFS measurement. However, we are rather convinced that the anomalous x-ray scattering technique is much more straightforward, at least theoretically, and the environmental structural information including so-called middle range ordering may be evaluated as a function of radial distance with much higher relia- bility than the EXAFS method. The following comment may also be suggested. The EXAFS method is undoubtedly one of the most powerful methods for determining the fine structure of various materials. However, as already mentioned by Lee et al.(1981), the EXAFS method is not easy to differentiate between a reduction in the short range order parameter and the degree of disorder, unless a considerable amount of fundamental structural information is already known about the desired materials. Then it is **unrealistic** to apply to a completely unknown and complicated material and expect **the EXAFS method alone** to provide the correct structural information. Thus the anomalous x-ray scattering should become, in the near future, a most reliable and powerful tool for studying the CSRO in multi-component disordered materials and could make a significant impact on the knowledge of their fine struc- tures. In addition, the anomalous x-ray scattering can be applicable to various disordered materials with only a few exceptions such as light elements. This advantage contrasts with other techniques such as neutron diffraction using anomalous scattering or isotope substitution technique. It may be also suggested that from considering many factors that various questions un-solved by conventional diffraction techni- ques and source of x-rays are almost undoubtedly answered by making available accurate partial structural functions or environmental structural information using this relatively new technique, when coupled with a synchrotron radiation source.

CHAPTER 2
FUNDAMENTAL RELATIONSHIPS BETWEEN RDF AND SCATTERING INTENSITY

All atomic positions in crystalline materials are well-described with a few parameters of distances and angles. However, such a simple definition is impossible in disordered materials, because of the lack of long-range structural periodicities. The structure of disordered materials can only be quantitatively described in terms of the so-called **radial distribution function (RDF)** indicating the average probability of finding another atom in a specified volume from an origin atom as a function of radial distance. The RDF gives spherically averaged information on the atomic correlation as one-dimensional data, however it does give almost undoubtedly unique quantitative information for describing the atomic scale structure of disordered materials. The description for the principles and their utility of RDF has already been provided in detail (see for example, Hill, 1956, Warren 1969, Wagner 1978), so that we give here only the essential points of the RDF analysis of disordered materials for convenience of discussion in this article.

In the case of an hypothetical homogeneous disordered system both over time and space, the so-called radial distribution function, RDF = $4\pi r^2 \rho(r)$, may be defined by considering a spherical shell of radius r with thickness dr centered on an origin atom. The quantity of $\rho(r)$ is often referred to as the radial density function which corresponds to the average probability of finding another atom as a function of only distance $\vec{r} = |\vec{r}' - \vec{r}_0|$. As shown in the schematic diagram of **Fig.2.1**, the RDF gradually approaches the parabolic function of $4\pi r^2 \rho_0$ at a larger value of r, where ρ_0 is the average number density of atoms, because the positional correlation of atoms in disordered materials weakens with increasing distance. Of course, no atomic positions exist within the minimum nearest neighbor distance such as the atomic core diameter due to the repulsion in the pair potential and the RDF is equal to zero in such small value of r. The area under the respective peak in the RDF yields information about the coordination number. The reduced radial distribution function of G(r) expressed by the following equation is also widely used for discussion about the atomic scale structure of disordered materials.

$$G(r) = 4\pi r [\rho(r) - \rho_0] \tag{2.1}$$

Another function of $g(r) = \rho(r)/\rho_0$, referred to as the pair distribution function, is also frequently used. It may be noted that the function of $g(r)$ is sometimes named the radial distribution function too, because of the function of radial distance r alone in disordered materials.

The RDF can be determined from diffraction data with x-rays, neutrons and electrons, because even in the disordered system with the lack of long-range periodicity, two atoms whose scattered beams coherently interfere with each other, result in the scattering intensity depending upon the relative positions of the two atoms. X-ray diffraction measurement is the most familiar and the most important method in structural analysis for disordered materials. Therefore, the x-ray case is presented here as an example, although many of the concepts and procedures are similarly applicable to the measurements by using neutron and electron diffractions.

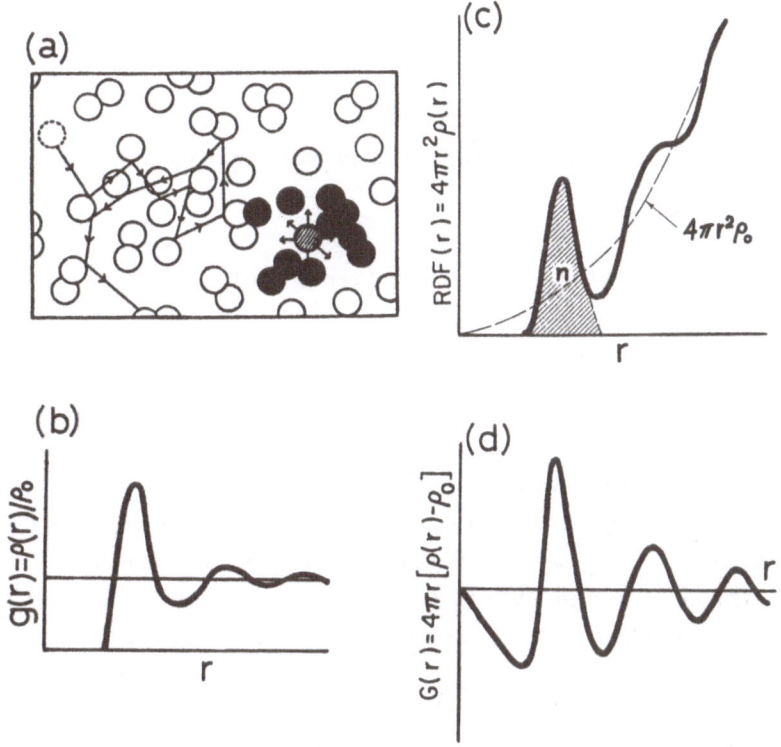

Fig.2.1 Schematic diagram of atomic distribution and radial distribution functions [$g(r)$, $4\pi r^2\rho(r)$ and $G(r)$] in the disordered state.

Writing that wave vector of incident and scattered x-rays as $\vec{q}_o$ and $\vec{q}'$, the diffraction vector q is defined as $\vec{q} = |\vec{q}' - \vec{q}_o|$ and its magnitude is expressed by;

$$q = |\vec{q}| = 4\pi\sin\theta/\lambda \qquad (2.2)$$

$$= \frac{4}{hc^o} \sin \cdot E \qquad (2.3)$$

where θ is half the scattering angle, λ is the wavelength of the incident x-rays, h and c^o are the Planck constant and the speed of light and E is the energy of the incident x-ray photon. Equation (2.3) is convenient for variable wavelength measurements such as energy dispersive x-ray diffraction (EDXD) method (see for example, Egami 1981b). Since the phase factor of scattered x-rays at the position of r is given by $\exp(-i\vec{q}.\vec{r})$, the amplitude of the scattered x-rays is expressed in the static approximation by the following form;

$$A(\vec{q}) = \sum_k f_k(q)\exp(-i\vec{q}\cdot\vec{r}_k) \qquad (2.4)$$

where $f_k(q)$ is the atomic scattering factor for atom k located at position of $\vec{r}_k$. Thus the coherent x-ray scattering intensity $I^{coh}(q)$ can be written by;

$$I^{coh}(q) = <A(\vec{q})A*(\vec{q})> = <\sum_{jk}f_j(q)f_k(q)\exp\{-i\vec{q}\cdot(\vec{r}_j-\vec{r}_k)\}> \qquad (2.5)$$

Here the brackets $< >$ denote the statistical average. In homogeneous disordered materials the long-range atomic periodicities known in crystals disappears, so that the summation in eq.(2.5) may be well-expressed by the average value of the positional correlation over all orientations. Due to the spherical symmetry, the functions f(q) and $I^{coh}(q)$ depend only upon the magnitude of the diffraction vector q (see for example, Warren 1969). It is noted, however, that the mono-tonic decrease in the atomic scattering factor f(q) with increasing q is attributed to the intra-atomic interference effect which is almost independent of the atomic distribution within the scattering process. Therefore, the coherent x-ray scattering intensity is normalized by removing this q-dependence from the f(q) term and then the structure factor, which is directly related to the RDF in disordered materials, can be defined as follows (see for example, Wagner 1978, Waseda 1980).

For simplicity consider at first a disordered system containing only one kind of atom. Equation (2.5) reduces to the following form in a one-component disordered materials.

$$I^{coh}(q) = f^2(q) < \sum_{jk} exp\{-i\vec{q} \cdot (\vec{r}_j - \vec{r}_r)\} >$$ (2.6)

Excluding the forward scattering term, the structure factor $S(q)$, often referred to the interference function, can be written by;

$$S(q) = \frac{1}{N} < \sum_{jk} exp\{-iq \cdot (r_j - r_k)\} > - N\delta_{\vec{q},o}$$ (2.7)

where N is the total number of atoms in the disordered system with volume V and $\delta_{\vec{q},0}$ term corresponds to the intensity at $q = 0$. The contribution from the $\delta_{\vec{q},0}$ term is frequently neglected in practical calculation, because its physical significance is limited to an extremely narrow region near $q = 0$.

On the other hand, the so-called radial density function $\rho(r)$ expressed by the following;

$$\rho(r) = \rho_o g(r) = \frac{1}{N} < \sum_{jk} \delta\{\vec{r} - (\vec{r}_j - \vec{r}_k)\} > - \delta(\vec{r})$$ (2.8)

By using the relation of $\rho(r) = \rho_o[g(r)-1] + \rho_o$, eq.(2.8) is then rewritten by;

$$\frac{1}{N} < \sum\sum \delta\{\vec{r} - (\vec{r}_j - \vec{r}_k)\} > - \rho_o = \rho_o[g(r)-1] + \delta(r)$$ (2.9)

We can obtain the following equation by applying the Fourier transformation to eq.(2.9);

$$\frac{1}{N} < \sum\sum exp\{-iq \cdot (\vec{r}_j - \vec{r}_k)\} > - N\delta_{\vec{q},o} = 1 + \rho_o \int [g(r)-1] exp(-i\vec{q} \cdot \vec{r}) d\vec{r}$$ (2.10)

Here, the following relations have also been used;

$$\frac{1}{V} \int exp(-i\vec{q} \cdot \vec{r}) d\vec{r} = \delta_{\vec{q},o}$$ (2.11)

$$\int \delta(r) exp(-i\vec{q} \cdot \vec{r}) dr = 1$$ (2.12)

$$\int \delta(\vec{r}-\vec{r}')\exp(-i\vec{q}\cdot\vec{r}')d\vec{r}' = \exp(-i\vec{q}\cdot\vec{r}) \qquad (2.13)$$

Thus, we can now get the following fundamental relation between the structure factor obtained by diffraction experiments and the radial distribution function for disordered materials.

$$S(q) = 1 + \rho_o \int \left[g(r)-1\right]\exp(-i\vec{q}\cdot\vec{r})d\vec{r} \qquad (2.14)$$

Of course, the following well-known equation for estimation of RDF is easily derived from eq.(2.14).

$$G(r) = 4\pi r\left[\rho(r)-\rho_o\right] = \frac{2}{\pi}\int_0^\infty q\left[S(q)-1\right]\sin(q\cdot r)dq \qquad (2.15)$$

On the other hand, the RDF analysis and its interpretation are more complicated in the case of disordered materials containing more than two kinds of atoms. However, when we introduce the compositionally averaged functions expressed by the following equations, a similar manner to that for the simple one-component case may be applicable as follows.

$$\bar{\rho}(r) = \sum_{i=1}^{n}\sum_{j=1}^{n} c_i f_i f_j \rho_{ij}(r)/<f>^2 \qquad (2.16)$$

$$<f>^2 = (\sum_{i=1}^{n} c_i f_i)^2 \qquad (2.17)$$

$$<f^2> = \sum_{i=1}^{n} c_i f_i^2 \qquad (2.18)$$

where c_i is the atomic fraction of i-type atom and $\rho_{ij}(r)$, generally called the partial radial density function, corresponds to the number of i-type atoms found at a radial distance of r from a j-type atom at the origin. Here q-dependence of the atomic scattering factor $f(q)$ is excluded for simplification. Equation (2.16) implies that the average radial density function $\bar{\rho}(r)$ of multi-component disordered materials could be given by the summation of the partial radial density function $\rho_{ij}(r)$ with a weighting factor using the atomic scattering factor and concentration. By using the approximate equations of eqs.(2.16)-(2.18), the coherent x-ray scattering intensity per atom $I_a^{coh}(q)$ and the structure factor a(q), often called total structure factor, for disordered materials containing more than two kinds of atoms are given by the following equations;

$$I_a^{coh}(q) = <f^2> + <f>^2 \int_0^\infty 4\pi r^2 [\rho(r) - \rho_o] \frac{\sin(q \cdot r)}{q \cdot r} \, dr \qquad (2.19)$$

$$a(q) = \left[I_a^{coh}(q) - (<f^2> - <f>^2) \right] / <f>^2 \qquad (2.20)$$

Thus, the following common relation, similar to eq.(2.15), may be obtained.

$$\bar{G}(r) = 4\pi r \left[\bar{\rho}(r) - \rho_o \right] = \frac{2}{\pi} \int_0^\infty q \left[a(q) - 1 \right] \sin(q \cdot r) \, dq \qquad (2.21)$$

Equation (2.21) gives the fundamental equation for extracting the information about the atomic scale structure from the observed x-ray scattering intensity data for disordered materials including more than two kinds of atoms, although the information of G(r) cannot be used to describe completely the positions and chemical identities of atoms. For this purpose, the knowledge of the structure of individual pairs like $\rho_{ij}(r)$ is undoubtedly required. Such information is often called partial structural functions and they are probably only unique items for understanding various characterisitic properties of multi-component disordered materials at a microscopic level.

CHAPTER 3

DEFINITION OF PARTIAL STRUCTURE FACTORS AND COMPOSITIONAL SHORT RANGE ORDER (CSRO)

Since the surrounding of each atom in disordered materials are generally different from those of other atoms as easily seen in Fig.2.1(a), the interpretation of RDF for the multi-component case is more complicated. In this respect, the partial RDFs for the individual pairs of chemical constituents are of particular importance and almost the only unique items for quantitatively describing the atomic scale structure of multi-component disordered materials. The partial radial density function $\rho_{ij}(r)$ can mathematically defined by,

$$\rho_{ij}(r) = c_j \rho_o g_{ij}(r) = \frac{1}{N_i} \sum_{i\alpha}^{N_i} \sum_{\beta}^{N_j} \delta\{\vec{r} - (\vec{r}_\alpha - \vec{r}_\beta)\} - \delta_{\alpha\beta}\delta(\vec{r}) \qquad (3.1)$$

where N_i is the number of i-type atoms in the volume V and $g_{ij}(r)$ is the so-called partial pair distribution function. The summation of i and j over the chemical constituents whereas the summation of α and β is over all atoms in the system. The following relations have also been used.

$$c_i \rho_{ij}(r) = c_j \rho_{ji}(r) \qquad (3.2)$$

$$N_i \rho_{ij}(r) = N_j \rho_{ji}(r) \qquad (3.3)$$

Therefore it follows immediately that $g_{ij}(r) = g_{ji}(r)$. Since the long-range atomic correlations disappear in disordered materials such as liquids and amorphous solids, the probability of finding a pair of atoms, of course, approaches the average value with an increase in distance. Thus,

$$\rho_{ij}(r) \rightarrow \frac{N_j}{V} = c_j \rho_o \qquad g_{ij}(r) \rightarrow 1 \qquad (r \rightarrow \infty) \qquad (3.4)$$

Hence, the basic features of the partial RDFs themselves are very similar to those of the RDF for the simple case containing only one kind of atoms. However, the number of partial RDFs, corresponding to the possible atomic pairs, drastically increases, as the number of constituent atoms in the system increases. It follows that there are $n(n+1)/2$ possible pairs in the system containing n-components and thus three partial RDFs in a binary system and six partial RDFs in a ternary system are required to describe completely the atomic scale structure in disordered materials.

The partial RDF is known to connect with the partial structure factor (often called partial interference function) of corresponding pair correlation determined from the diffraction experiments of x-rays and neutrons. However, the definition of the partial structure factors are not unique and three different equations have been used in the literature, although they are combined by linear relations (see for example, Waseda 1980). For convenience, the essential points of these different sets of the partial structure factors are given below, using a binary disordered system as an example.

Let us consider a binary disordered material containing two types of atoms, 1 and 2. Thus three different partial RDFs, related to $\rho_{11}(r)$, $\rho_{22}(r)$ and $\rho_{12}(r)$, are required to describe this structure. The total number of atoms N in a volume V consists of N_1 and N_2 where N_1 and N_2 are the number of atoms 1 and 2 and then the atomic fractions are defined by $c_1 = N_1/N$ and $c_2 = N_2/N$, respectively. The coherent x-ray scattering intensity $I^{coh}(q)$, analogous to eq.(2.5) may be written by the following form.

$$I^{coh}(q) = f_1^2 < |\Sigma \exp(-i\vec{q}\cdot\vec{r}_{ij})|^2 > + f_2^2 < |\sum_{k=1}^{N_2}\exp(-i\vec{q}\cdot\vec{r}_{2k})|^2 >$$

$$+ 2f_1f_2 < \sum_{j=1}^{N_1}\sum_{k=1}^{N_2} \exp\{-i\vec{q}\cdot(\vec{r}_{ij}-\vec{r}_{2k})\} > \qquad (3.5)$$

Here, the q-dependence of the atomic scattering factor $f_i(q)$ is excluded for simplification, because it depends only upon the intra-atomic interference effect. The three double sums in eq.(3.5) corresponds to the partial structure factors of the respective atomic pairs, 1-1, 2-2 and 1-2.

Faber and Ziman (1965) used the following definition, analogous to eq.(2.14), for the partial structure factors $a_{ij}(q)$;

$$a_{ij}(q) = 1 + \rho_o \int [g_{ij}(r)-1] \exp(-i\vec{q}\cdot\vec{r})d\vec{r} \qquad (3.6)$$

Excluding the forward scattering term, the Faber-Ziman (hereafter referred to as FZ) partial structure factors can also be written by;

$$a_{ij}(q) = (c_i c_j)^{-1/2} \left[(N_i N_j)^{-1/2} < \sum_{\alpha} \sum_{\beta} \{-i\vec{q} \cdot (\vec{r}_{i\alpha} - \vec{r}_{j\beta})\} > \right.$$

$$\left. - (N_i N_j)^{1/2} \delta_{\vec{q},o} - c_j^{-1} \delta_{ij} + 1 \right] \qquad (3.7)$$

By using the FZ formulation, the coherent x-ray scattering intensity per atom is given as follows;

$$I_a^{coh}(q) = (<f^2> - <f>^2) + \sum_{ij} c_i c_j f_i f_j a_{ij}(q) \qquad (3.8)$$

where

$$<f^2> = c_1 f_1^2 + c_2 f_2^2 \qquad (3.9)$$

$$<f> = c_1 f_1 + c_2 f_2 \qquad (3.10)$$

$$(<f^2> - <f>^2) = c_1 c_2 (f_1 - f_2)^2 \qquad (3.11)$$

The quantity given by eq.(3.11) is frequently called the **Laue monotonic scattering term** attributed to the intensity arising only from the difference in the atomic scattering factors of the constituent atoms. The FZ total structure factor is then expressed by the following equation.

$$a_{FZ} = \left[I_a^{coh}(q) - (<f^2> - <f>^2) \right] / <f>^2 \qquad (3.12)$$

$$= \sum_{ij} c_i c_j \frac{f_i f_j}{<f>^2} a_{ij}(q) \qquad (3.13)$$

These FZ expressions are used in eqs.(2.19) and (2.20).

On the other hand, Ashcroft and Langreth (1967) (hereafter referred to as AL) used a different expression for the partial structure factors $S_{ij}(q)$. Their definitions corresponding to the FZ case are as follows:

$$S_{ij}(q) = \delta_{ij} + (c_i c_j)^{1/2} \rho_o \int \left[g_{ij}(r) - 1 \right] \exp(-i\vec{q}\cdot\vec{r}) d\vec{r} \qquad (3.14)$$

$$S_{ij}(q) = (N_i N_j)^{-1/2} \langle \sum_{\alpha\beta} \exp\{-i\vec{q}\cdot(\vec{r}_{i\alpha} - \vec{r}_{j\beta})\} \rangle - (N_i N_j)^{1/2} \delta_{\vec{q},o} \qquad (3.15)$$

$$I_a^{coh}(q) = \sum_{ij} (c_i c_j)^{1/2} f_i f_j S_{ij}(q) \qquad (3.16)$$

$$S_{AL}(q) = I_a^{coh}(q) / \langle f^2 \rangle \qquad (3.17)$$

$$= \sum_{ij} (c_i c_j)^{1/2} \frac{f_i f_j}{\langle f^2 \rangle} S_{ij}(q) \qquad (3.18)$$

These different sets of partial structure factors can mutually trans-formed by the following linear equations.

$$\left. \begin{array}{l} S_{11}(q) = 1 + c_1 \left[a_{11}(q) - 1 \right] \\[2mm] S_{22}(q) = 1 + c_2 \left[a_{22}(q) - 1 \right] \\[2mm] S_{12}(q) = (c_1 c_2)^{1/2} \left[a_{12}(q) - 1 \right] \end{array} \right\} \qquad (3.19)$$

A brief comment is also given below regarding the physical significance of the FZ and AL partial structure factors. As easily seen in eqs.(3.13) and (3.18), different normalizations have to be done. The three partial structure factors in the FZ form vary around unity, while the AL partial structure factor of unlike atom pairs $S_{12}(q)$ oscillates around zero in contrast to the variation around unity in the two partial structure factors of like atom pairs, $S_{11}(q)$ and $S_{22}(q)$. The FZ form expresses the mutually comparable quantities, i.e., for a substitutional alloy system in which a solute atom can replace a solvent atom without any constraint such as the change in volume. The concentration dependence of the FZ partial structure factors corresponds to the deviation from ideal behavior where the concentration independence is well recognized (see for example, Halder and Wagner 1967). This is the reason for the fact that the FZ partial structure factors (eq.(3.6)) are relatively insensitive to the concen-tration compared with the AL partial structure factors (eq.(3.14)). However, it is worth mentioning that the partial structure factors of both definitions should, in principle, be function of the concentration. Using eqs.(3.13) and (3.18), the following relation could be made regarding the total structure factors:

$$\frac{S_{AL}(q) - 1}{a_{FZ}(q) - 1} = \frac{<f>^2}{<f^2>} = 1 - \frac{c_1 c_2 (f_1 - f_2)^2}{<f^2>} \tag{3.20}$$

Thus,

$$
\left.
\begin{aligned}
S_{AL}(q) &< a_{FZ}(q) & &\text{when} & a_{FZ}(q) &> 1 \\[2ex]
S_{AL}(q) &= a_{FZ}(q) = 1 & &\text{when} & a_{FZ}(q) &= 1 \\[2ex]
S_{AL}(q) &> a_{FZ}(q) & &\text{when} & a_{FZ}(q) &< 1
\end{aligned}
\right\} \tag{3.21}
$$

An additional set of partial structure factors were proposed by Bhatia and Thornton (1970). Their approach provides a useful link between the mean square thermal fluctuation in local number density and concentration in disordered materials on the one hand and macroscopic thermodynamic properties on the other, which is strictly correct only in the long wavelength limit ($q \to 0$). However, the Bhatia-Thornton (hereafter referred to as BT) partial structure factors themselves correspond to the Fourier transforms of the mean square fluctuations in the particle number, the concentration and their cross term. This formalism is superior to the FZ and AL forms, when discussing several specific topics of disordered materials such as the relation between structure and thermodynamic quantities or some kinds of atomic ordering effect arising from different physico-chemical properties of constituent atoms in multi-component system. The detailed description for the BT partial structure factors and their physical significance in the long wavelength limit are given in **Appendix 1** for future reference. However, we give here some essential points of the Bhatia and Thornton's expression for comparison with the FZ or AL form.

Bhatia and Thornton (1970) proposed three partial structure factors $S_{NN}(q)$, $S_{CC}(q)$ and $S_{NC}(q)$ using the concept of the local fluctuations in disordered materials. $S_{NN}(q)$ is the spatial correlation between number density (related to the topological arrangement of atoms) and its Fourier transform corresponds to the usual RDF. $S_{CC}(q)$ is the correlation of the concentration fluctuation for chemical constituents and its Fourier transform gives the so-called **radial concen-**

tration correlation function (RCF) named by Ruppersberg and Egger (1975). The cross term of $S_{NC}(q)$ contains information about concentration fluctuations which may be stimulated by the topological effect of the constituent atoms (and vice-versa) and thus this cross term is likely to be the size effect. Then the Fourier transform of $S_{NC}(q)$ corresponds to the radial function of the size effect including the concentration fluctuation in disordered materials. The BT partial structure factors are given by the following equations, analogous to eq. (3.6);

$$S_{NN}(q) = 1 + \rho_0 \int \left[g_{NN}(r) - 1 \right] \exp(-i\vec{q}\cdot\vec{r}) d\vec{r} \tag{3.22}$$

$$S_{CC}(q) = c_1 c_2 + \rho_0 \int g_{NC}(r) \exp(-i\vec{q}\cdot\vec{r}) d\vec{r} \tag{3.23}$$

$$S_{NC}(q) = \rho_0 \int g_{NC}(r) \exp(-i\vec{q}\cdot\vec{r}) d\vec{r} \tag{3.24}$$

The expressions such as $G_{CC}(r) = 4\pi r \rho_0 g_{CC}(r)$ or $4\pi r^2 \rho_{CC}(r)$ are also accepted in manner similar to the previous discussion. By using this new set of partial structure factors, the coherent x-ray scattering intensity per atom $I_a^{coh}(q)$ and the BT total structure factor $S_{BT}(q)$ can be expressed as follows.

$$I_a^{coh}(q) = \langle f^2 \rangle S_{NN}(q) + \Delta f^2 S_{CC}(q) + 2\langle f \rangle \Delta f S_{NC}(q) \tag{3.25}$$

$$S_{BT}(q) = I_a^{coh}(q)/\langle f^2 \rangle \tag{3.26}$$

$$= \frac{\langle f \rangle^2}{\langle f^2 \rangle} S_{NN}(q) + \frac{\Delta f^2}{\langle f^2 \rangle} S_{CC}(q) + 2 \frac{\langle f \rangle \Delta f}{\langle f^2 \rangle} S_{NC}(q) \tag{3.27}$$

where $\Delta f = f_1 - f_2$. The BT partial structure factors are also combined by linear equations to both the FZ and AL partial structure functions. They are given below for convenience.

$$S_{NN}(q) = c_1^2 a_{11}(q) + c_2^2 a_{22}(q) + 2c_1 c_2 a_{12}(q)$$

$$S_{CC}(q) = c_1 c_2 \left[1 + c_1 c_2 \{ a_{12}(q) + a_{22}(q) - 2a_{12}(q) \} \right]$$

$$\left.\begin{array}{c} \end{array}\right\} \tag{3.28}$$

$$S_{NC}(q) = c_1 c_2 \left[c_1 \{ a_{11}(q) - a_{12}(q) \} - c_2 \{ a_{22}(q) - a_{12}(q) \} \right]$$

$$S_{NN}(q) = c_1 S_{11}(q) + c_2 S_{22}(q) + 2(c_1 c_2)^{1/2} S_{12}(q)$$

$$S_{CC}(q) = c_1 c_2 \left[c_2 S_{11}(q) + c_1 S_{22}(q) - 2(c_1 c_2)^{1/2} S_{12}(q) \right] \qquad (3.29)$$

$$S_{NC}(q) = c_1 c_2 \left[S_{11}(q) - S_{22}(q) + (c_2 - c_1) S_{12}(q) / (c_1 c_2)^{1/2} \right]$$

The following points on the BT structure factors may be suggested. With respect to the total structure factor, $S_{BT}(q)$ is rather
similar to the $S_{AL}(q)$ case than that of $a_{FZ}(q)$. $S_{NN}(q)$ oscillates
around unity, whereas $S_{CC}(q)$ oscillates around the value of the
corresponding average concentration $c_1 c_2$ in a binary disordered
material and thus the expression of $S_{CC}(q)/c_1 c_2$ has also been used in
the literature. The cross term of $S_{NC}(q)$ oscillates around zero and
disappears for random mixtures, i.e., mixtures in which the local
number density of i and j atoms around a i-type atom is the same as
that around a j-type one. In addition, $S_{CC}(q)$ is independent of q for
random mixtures and its constant value is $c_1 c_2$. This leads to the fact
that the term of $\Delta f^2 S_{CC}(q)$ in eq.(3.25) corresponds to the Laue monotonic scattering term, $<f^2> - <f>^2$ given in eq.(3.11) and then the
relation of $<f>^2 S_{NN}(q) = \Sigma \Sigma c_i c_j f_i f_j a_{ij}(q)$ can be deduced from
eqs.(3.8) and (3.25) for random mixtures. It is also noted that $S_{NN}(q)$
$\rightarrow 1$, $S_{CC}(q) \rightarrow c_1 c_2$ and $S_{NC}(q) \rightarrow 0$, when $q \rightarrow \infty$. If the unlike atom pair
correlation is dominant rather than the like atom pair correlations;
$S_{CC}(q) < c_1 c_2$ and then $g_{CC}(r)$ shows the negative value;while when the
like atom pair correlations are superior to that of the unlike atom
pairs; $S_{CC}(q) > c_1 c_2$ and $g_{CC}(r)$ gives the positive value.

The structural features of disordered materials should be characterized by the atomic short-range order directly related to the RDF.
However, anyone who studies disordered materials, also knows the following points. The characteristic features of respective crystal
structure (long-range atomic periodicity) disappear in the disordered
state and the increase in the freedom of atomic configurations due to
the increase of the vacant space contributes to the construction of
universal short-range ordering and then the strucrtural functions,
such as RDF, of various disordered materials look similar (Waseda
1980). This results in the fact that their basic structural features
could be explained more or less by the model of **dense random packing
of hard spheres**(see for example, Bernal 1959, Cargill III 1975). For
this reason, there is still some confusion about the concept of atomic

short-range order and no unique definition is available for describing completely the atomic short-range order in disordered materials at the present time. A completely random mixing is also not obtained in most cases, particularly in the disordered state of condensed matter, so that the distribution of constituent atoms around the atoms of each component likely differs from the average value. Such deviation from simple mixtures probably approximated by the hard sphere mixtures of different sizes (see for example, Ashcroft and Langreth 1967) is usually conceivable, when disordered materials consists of more than two kinds of atoms having different physico-chemical properties such as size, charge number and electronegativity. This deviation from the simple average indeed corresponds to the so-called **compositional short range order** (hereafter referred to as **CSRO**) in multi-component disordered materials.

It may be convenient from this point of view if the idea of the **Warren chemical short range order parameter,** originally proposed for crystalline substitutional solid solutions (see for example, Warren 1969), is widely extended to disordered materials. The Warren chemical short range order parameter of the p-th coordination shell α_p for a binary system is given by the following equation in terms of the RDFs and the concentration.

$$\alpha_P = 1 - \frac{\rho_{12}(r)}{\rho(r)c_2} = 1 - \frac{\rho_{21}(r)}{\rho(r)c_1} \qquad (3.30)$$

The atoms in disordered materials no longer occupy the fixed position in a periodic three-dimensional lattice and therefore the evaluation of α_p is not as straightforward as in crystalline materials. However, this method is one way to provide information about the preference for unlike atom pairs or for like atom pairs in disordered materials.

The effect of CSRO on the properties of disordered materials is feasibly related to, in most cases, the first nearest neighbor atomic correlations, on the basis of various experimental data (see for example, Faber 1972, Egami 1981a). From this point of view, a crude but useful parameter for indicating CSRO in disordered materials may be given only by the deviation in the nearest neighbor atomic correlations from the average. Such parameter for CSRO in a binary disordered system can be simply expressed in terms of the coordination numbers.

$$\alpha_p = 1 - \frac{n_{12}}{\langle n \rangle c_2} = 1 - \frac{n_{21}}{\langle n \rangle c_1} \qquad (3.31)$$

where n_{ij} is the nearest coordination number of j-type atoms around a i-type atom, $\langle n \rangle$ is the total coordination number in the nearest neighbor region and c_i denotes the concentration of i-type atom. This CSRO parameter is now known as the **Warren-Cowley short range order parameter** (see for example, Warren et al 1951, Cowley 1950) and more detailed discussion for this short range order parameter of α_p has already been available for disordered materials (Wagner 1978, Cargill III and Spaepen 1981). We can estimate both coordination numbers, n_{12} and $\langle n \rangle$, from RDF data by using the following equations.

$$n_{12} = \int_{r_p'}^{r_p''} 4\pi r^2 \rho_{12}(r)dr \qquad (3.32)$$

$$\langle n \rangle = \int_{r_p'}^{r_p''} 4\pi r^2 \rho(r)dr = \int_{r_p'}^{r_p''} 4\pi r^2 \rho_{NN}(r)dr \qquad (3.33)$$

where r_p' and r_p'' are the left hand edge and the right hand edge of the corresponding peak in the respective RDF curve, respectively. When the radial concentration correlation function, $4\pi r^2 \rho_{CC}(r)$, is available, the following equation is also used for evaluation of the CSRO parameter as first applied by Ruppersberg and Egger (1975) for condensed disordered materials.

$$\alpha_p \langle n \rangle = \int_{r_p'}^{r_p''} 4\pi r^2 \rho_{CC}(r)dr \qquad (3.34)$$

The CSRO parameter expressed by eqs.(3.30) or (3.31) provides a useful measure of chemical affinity in disordered materials as long as the size difference of the constituent atoms is not too large. For example, if a preference of unlike atom pairs exists ($\rho_{12}(r) > \rho(r)c_2$ and then $n_{12} > \langle n \rangle c_2$), α_p shows a negative value, whereas a positive value of α_p implies that the correlation of like atom pairs ($\rho_{12}(r) < \rho(r)c_2$ and then $n_{12} < \langle n \rangle c_2$) is preferred to that of unlike atom pairs. It should be kept in mind, however, that the determination of CSRO as a function of distance is required for more involved studies on the structure-property relationships for disordered materials. For this purpose, the radial concentration function of RCF = $4\pi r^2 \rho_{CC}(r)$, in the BT form is almost undoubtedly the best information. We may also add that some further consideration such as the directional CSRO will also be required for explaining several properties exemplified by the magnetic anisotropy frequently found in metallic glasses (see for example, Graham Jr and Egami 1978), because the present information given by RDF including RCF corresponds to the spherically averaged data (nondirectional).

CHAPTER 4

EXPERIMENTAL DETERMINATION OF PARTIAL STRUCTURAL FUNCTIONS

Most of the important disordered materials such as metallic glasses and amorphous semiconductors are known to contain more than two kinds of atoms and then only a weighted sum of the partial structure factors, which represent the Fourier transform of the partial RDFs, can be obtained from a single diffraction experiment with x-rays, neutrons and electrons. The concept and utility of the partial structural functions have been emphasized for a long time (see for example, Vineyard 1958, Keating 1963) and when the full set of the partial structural functions are obtained we can have more than just one-dimensional information and enables us to understand the structure-property relationships of disordered materials in a more realistic way including the compositional short range order (CSRO). Therefore, the determination of partial structural functions is almost undoubtedly one of the most important research subjects for disordered materials. As mentioned previously, the most obvious is the efficient application of the anomalous (resonance) x-ray scattering technique allowing the exploitation of a relatively new and powerful method for this purpose when coupled with the intense white x-ray source such as synchrotron radiation. We give here some essential points for separating the so-called partial structure factors in a binary disordered system from measured scattering intensity data.

First, it may be helpful to recall some fundamental equations. The coherent x-ray scattering intensity per atom $I_a^{coh}(q)$ for a binary disordered system is expressed by the following equation in terms of the Faber-Ziman (FZ) form.

$$I_a^{coh}(q) = (<f^2> - <f>^2) + <f>^2 a(q) \qquad (4.1)$$

where $a(q)$ is the total structure factor and the subscript of FZ used in eq.(3.12) is omitted here for simplification. Using the Fourier transformation, we obtain the total RDF as follows.

$$4\pi r^2 [\rho(r) - \rho_o] = rG(r) = \frac{2r}{\pi} \int_0^\infty q[a(q)-1] \sin(q \cdot r) dq \qquad (4.2)$$

The total structure factor a(q) corresponding to the structurally sensitive part in eq.(4.1) is characterized by the three partial structure factors $a_{ij}(q)$ in the following form.

$$a(q) = w_{11}a_{11}(q) + w_{22}a_{22}(q) + 2w_{12}a_{12}(q) \qquad (4.3)$$

$$w_{ij} = c_ic_jf_if_j/<f>^2 \qquad (4.4)$$

where c and f are the concentration and the atomic scattering factor, respectively. The partial structure factors $a_{ij}(q)$ correspond to the Fourier transform of the partial RDFs. These quantities are expressed by the following equations.

$$a_{ij}(q) = 1 + \int_0^\infty 4\pi r^2 \rho_o [g_{ij}(r)-1]\frac{\sin(q \cdot r)}{q\ r}dr \qquad (4.5)$$

$$4\pi r^2 \rho_o[g_{ij}(r)-1] = rG_{ij}(r) = \frac{2r}{\pi}\int_0^\infty q[a_{ij}(q)-1]\sin(q \cdot r)dq \qquad (4.6)$$

where $g_{ij}(r) = \rho_{ij}(r)/(c_j\rho_o)$. Regarding the partial RDFs, it is also customary to use the following relation.

$$G(r) = w_{11}G_{11}(r) + w_{22}G_{22}(r) + 2w_{12}G_{12}(r) \qquad (4.7)$$

The separation of three individual structural functions ($a_{ij}(q)$ or $G_{ij}(r)$) is the present objective. We will discuss here only the partial structure factors $a_{ij}(q)$, because $G_{ij}(r)$ is straightforward as given by the common Fourier transformation in the manner of eq.(4.6).

As it follows readily from eq.(4.3), the total structure factor a(q) of a binary disordered system containing atom 1 and 2 can be expressed by the summation of two like atom pairs (1-1 and 2-2) and one unlike atom pair (1-2), while the coefficient w_{ij}, frequently called weighting factor, depends on the concentrations and the atomic scattering factors. Thus the individual partial structure factors can be estimated only by making available at least three independent scattering experiments for which the weighting factors are varied without any change in the RDF. For example, when the scattering ability is changed in the component of 1, the following matrix form can be readily obtained.

$$
\begin{bmatrix}
a_{11}(q) \\[2ex]
a_{22}(q) \\[2ex]
a_{12}(q)
\end{bmatrix}
=
\begin{bmatrix}
c_1^2 f_1^2 & c_2^2 f_2^2 & 2c_1 c_2 f_1 f_2 \\[2ex]
c_1^2 (f_1')^2 & c_2^2 f_2^2 & 2c_1 c_2 f_1^* f_2 \\[2ex]
c_1^2 (f_1'')^2 & c_2^2 f_2^2 & 2c_1 c_2 f_1^{**} f_2
\end{bmatrix}^{-1}
\begin{bmatrix}
a(q) \\[2ex]
a^*(q) \\[2ex]
a^{**}(q)
\end{bmatrix}
\qquad (4.8)
$$

This can be done by several methods. Although they are not trivial tasks in practice, the following classification may be recognized in principle.

(A) the three different radiation technique using x-ray, neutron and electron diffractions.

(B) the isotope substitution technique for neutron diffraction in which the scattering powers of the components are varied by using different isotopes.

(C) the polarized neutron diffraction technique which is applicable only to magnetic materials.

(D) the anomalous scattering technique for both x-rays and neutrons.

Of course, an assortment of the above techniques such as the combination of x-ray and neutron diffraction experiments with the polarized neutron technique (Sadoc and Dixmier 1976) has also been used in the literature. **Table 4.1** shows the variation of these techniques for determining the partial structure factors in a binary disordered materials and the systems applied to.

The relative advantage and disadvantage of the respective diffraction technique have previously been discussed (Enderby 1968, Wagner 1978, Waseda 1980, 1981) and are not duplicated here. However, the following intrinsic point is considered to be worthy of note. The technique (A) requires different samples in size; bulk ($\simeq$ mm) for neutron, foil ($\simeq \mu$m) for x-rays and thin film ($\simeq$ nm) for electrons. In technique (B), the structure is automatically assumed to remain identical upon substitution by the isotopes. Thus, the fundamental idea of these two techniques (A) and (B) should be attributed to not only chemical but also structural identity. This assumption may be valid in a thermodynamically equilibrated disordered state such as in liquids. However, we should keep in mind that such structural identity may not remain in a thermodynamically metastable state such as glasses, because some properties of the as-quenched glasses particularly metal-

Table 4.1 Various methods for separating the partial structural functions in a binary disordered materials and their selected examples.

Combination of diffraction technique		Results
(A) x-ray, neutron and electron	glass	Pd-Si[1], Mn-Si[2]
(B) neutron (isotope substitution technique)	glass	Cu-Zr[3], Fe-Ge[4], TiO$_2$[5]
	liquid	Cu-Sn[6], Ag-Ge[7], NaCl[8],RbCl[9]
		CuCl[,10,11], AgCl[12], BaCl$_2$[13],
		ZnCl$_2$[14], CaCl$_2$[15], SrCl$_2$[16]
(C) neutron (polarized neutron technique)	glass	Co-P[17]
(D) x-ray + isotope substitution technique	glass	Fe-B[18], Ni-B[19], Ni-Nb[20]
	liquid	Cu-Sn[21]
(E) x-ray + polarized neutron technique	glass	Co-P[22], Tb-Fe[23],*
(F) x-ray anomalous scattering	glass	As-Te[24], Ni-P[25], Ni-Nb[26]
		Cu-Zr[26], Ge-Se[27]
	liquid	Ni-Si[28], Ce-Ni[29]
	solid electrolyte	α-Ag$_2$S[30], α-AgI[31]
(G) x-ray anomalous scattering + neutron	glass	GeO$_2$[32]

1. T.Masumoto, T.Fukunaga and K.Suzuki: Abstracts of Topical Conf. on the Atomic Scale Structure of Amorphous Solids, Yorktown Heights, New York (1978), p.F-7; Bull. Amer. Phys. Soc. 23(1978) 467; Sci. Rep. Res. Inst. Tohoku University, 28A(1980)208.

2. F.Paasche, H.Olbrich, G.Rainer-Harbach, P.Lamparter and S.Steeb: Z. Naturforsch. 37a(1982)1215.

3. T.Mizoguchi, T.Kudo, T.Irisawa, N.Watanabe, N.Niimura, M.Misawa and K.Suzuki: Proc. 3rd Inter. Conf. on Rapidly Quenched Metals, Brighton (1978), The Metals Society (London), Conf. Proc. No. 198, (1978), p.384; J. Phys. Soc. Japan 45(1978)1773.

4. K.Yamada, Y.Endoh, Y.Ishikawa and N.Watanabe: J. Phys. Soc. Japan 48(1980)922.

5. A.C.Wright and A.J.Leadbetter: Phys. Chem. Glasses 17(1976)122.

6. J.E.Enderby, P.A.Egelstaff and D.M.North: Phil. Mag. 14(1966)961.

7. M.C.Bellissent-Funel, P.J.Desre, R.Bellissent and G.Tourand: J. Phys. F. 7(1977)2485.

8. F.G.Edwards, J.E.Enderby, R.A.Howe and D.I.Page: J. Phys. C. 8(1975)3483.

9. E.W.J.Mitchell, P.F.J.Ponet and R.J.Stewart: Phil. Mag. 34(1976)721.

10. D.I.Page and K.Mika: J. Phys. C. 4(1971)3034.

11. S.Eisenberg, J.F.Jal, J.Dupuy, P.Chieux and W.Knoll: Phil. Mag. 46(1982)195.

12. J.Y.Derrin and J.Dupuy: Phys. Chem. Liquids 5(1976)71.

13. F.G.Edwards, R.A.Howe, J.E.Enderby and D.I.Page: J. Phys. C. 11(1978)1053.

14. S.Biggin and J.E.Enderby: J. Phys. C. 14(1981)703.

15. S.Biggin and J.E.Enderby: J. Phys. C. 11(1978)3577.

16. R.L.McGreevy and E.W.J.Mitchell: J. Phys C. 15(1982)5537.

17. J.Blétry and J.F.Sadoc: J. Phys. F. 5(1975)L110.

18. E.Nold, P.Lamparter, H.Olbrich, G.Rainer-Harbach and S.Steeb: Z. Naturforsch. 36a(1981)1032.

19. P.Lamparter, W.Sperl, S.Steeb and J.Blétry: Z. Naturforsch. 37a(1982)1223.

20. E.Sváb, F.Forgacs, F.Hajdu, N.Kroó and J.Takács: J. Non-Cryst. Solids 45(1978)1773.

21. Y.Waseda and K.Hirata: Bull. Res. Inst. Min. Met. Tohoku University, 31(1975)8.

22. J.F.Sadoc and J.Dixmier: Mater. Sci. Eng. 23(1976)187.

23. W.P.O'Leary: J. Phys. F. 5(1975)L175.

24. P.S.Gopalakvishnan and S.Ramaseshan: Acta Cryst. A31(1975)S159.

25. Y.Waseda and S.Tamaki: Z. Physik B23(1976)315.

26. H.S.Chen and Y.Waseda: phys. stat. sol. (a) 51(1979)593.

27. P.H.Fuoss, W.K.Warburton and A.Bienenstock: J. Non-Cryst. Solids 35/36(1980)1233.

28. Y.Waseda and S.Tamaki: Phil. Mag. 32(1975)951.

29. Y.Waseda and S.Tamaki: J. Phys. F. 7(1977)L151.

30. Y.Tsuchiya, S.Tamaki, Y.Waseda and J.M.Toguri: J. Phys. C. 11(1978)651.

31. Y.Tsuchiya, S.Tamaki and Y.Waseda: J. Phys. C. 12(1979)5361

32. P.Bondt: Acta Cryst. A30(1974)4,0.

* The isomorphous substitution technique (for example, Cargill and Tsuei 1978) was also used for Tb and Gd atoms.

lic glasses are known to vary from one run of production to the other and from one portion of the ribbon to another, clearly indicating the variation of CSRO at the microscopic level. On the other hand, the techniques (C) and (D) are free from this ambiguity, because it is possible to vary the weighting factors without the use of different samples.

Two additional methods, the concentration technique (see for example, Halder and Wagner 1967) and the isomorphous substitution technique (see for example, Chipman et al 1978, Cargill III and Tsuei 1978), have also been used to evaluate the partial structure factors in a binary or pseudo-binary disordered materials. Both techniques are possible ways with which to obtain some useful information on the partial structure factors in disordered materials in cases where other techniques are found to be technically difficult. However, note that their applicability strongly depends on the assumptions; concentration independence of the partial structures or structural identity of the samples in which at least one component is substituted by a chemically similar element. Thus a similar ambiguity to that of the techniques (A) and (B) holds true for these two optional techniques.

As mentioned in the previous section, three different definitions have been used for describing the partial structure factors in a binary disordered materials. The Bhatia-Thornton form is of particular interest in the case where $<f>$ might be zero (frequently called **zero alloy**). This can be made in neutron diffraction for samples containing an isotope with negative neutron cross section such as ^{7}Li and ^{62}Ni. The total structure factor for such a zero alloy contains only information of the partial structure factor of concentration-concentration correlation, $S_{CC}(q)$. However, it should be remembered that the complete set of the partial structure factors in the BT form, $S_{NN}(q)$, $S_{CC}(q)$ and $S_{NC}(q)$, can not be obtained only from the experiment for a zero alloy. Two additional independent scattering experiments are again required for this purpose, and when a full set of the partial structure factors in any one of these three forms is available, the partial structure factors in the other two forms can be readily evaluated by the linear equations in the manner of eqs.(3.28) and (3.29).

CHAPTER 5

NATURE OF ANOMALOUS X-RAY SCATTERING AND ITS APPLICATION FOR
STRUCTURAL ANALYSIS OF DISORDERED MATERIALS

It is well-known that the absorption edges for x-rays are found
in each atom at the characteristic energies and represent the
threshold excitation energies above which an inner electron can be
ejected into the continuum states (see for example, James 1954). These
absorption phenomena for x-ray include predominantly the excitation of
K-shell or L-shell electrons and thus our intention focuses mainly
upon the K-absorption edge or the L-absorption edge. In conventional
x-ray diffraction analysis, we generally choose the incident x-ray
energy (or wavelength) away from such an absorption edge of the constituent atoms and the energy independence is then well accepted for
the so-called atomic scattering factor f given by the simple potential
scattering theory. On the other hand, when the energy (or wavelength,
hereafter the term of energy is used) of the incident x-ray beam is
close to such an absorption edge of the constituent atoms, the atomic
scattering factor f becomes complex and can be expressed in the following form.

$$f(q,E) = f^o(q) + f'(E) + if''(E) \qquad (5.1)$$

where q and E are the wave vector and the incident x-ray energy,
respectively. The $f^o(q)$ corresponds to the normal atomic scattering
factor given by the Fourier transform of the electron density in atom,
for radiation at an energy much higher than any absorption edge and f'
and f" are the real and imaginary components of the so-called anomalous dispersion term (James 1954).

Since the spatial distribution of inner electrons is considerably
smaller than the magnitude of the x-ray wavelength, the dipole ap-
proxiamtion [$\exp(-i\vec{q}.\vec{r}) \simeq 1$] is well accepted and then the q-dependence
of the anomalous dispersion factors f' and f" can be ignored. Such q-
dependence may be required for discussion of f' and f" near the
absorption edges of M and N series (Wagenfeld 1975).

The anomalous dispersion factors f' and f" arising from anomalous
(resonance) scattering depend upon the incident energy of x-rays and
their variation for the energy is given in **Fig.5.1** using the K-
absorption edge of Ni atom as an example. The salient features are as

follows. The imaginary part of f" is positive and distinguished only on the higher energy side of the absorption edge. On the other hand, the real part of f' indicates a sharp negative peak at the absorption edge and its width is typically 50 eV at half maximum. The component f' exists on either side of the edge and the maximum effect at the energy rapidly reduces and approaches an approximately constant level for energies a few hundred eV away from the edge.

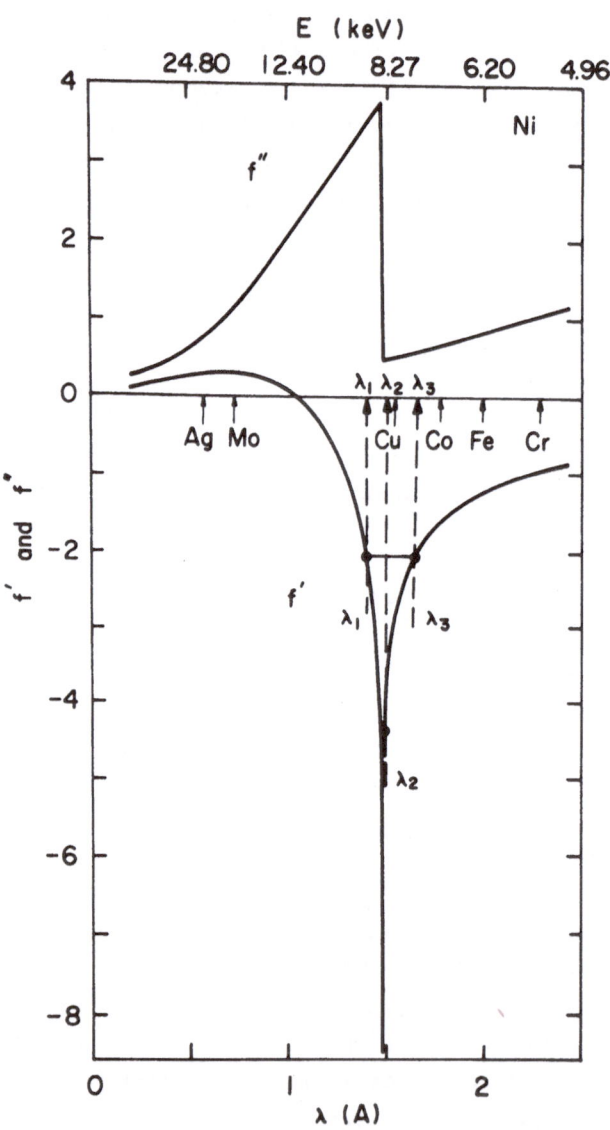

Fig.5.1 Energy (wavelength) dependence of anomalous dispersion factors f' and f" for Ni atoms calculated by the Cromer-Liberman's scheme.

The dispersion behavior of f" follows the simple absorption phenomena, i.e., the absorption edge corresponds to the threshold frequency (energy) above which an inner electron can be ejected into the continuum states and this process can take place only for the incident frequency equal to or greater than that of the absorption edge as easily seen in Fig.5.1. The real part of f' is generally observed as a phase difference in the x-ray optics and its dispersion behavior is not independent of the dispersion behavior of the imaginary part f". The energy dependence of f' corresponds to the so-called Kramers-Krönig transform of the imaginary component of f". This is frequently called **"dispersion relation"** and can be written in the following generalized form (see for example, James 1954);

$$f'(\omega) = \frac{2}{\pi} \int_0^\infty \frac{f''(\omega)\omega'}{\omega^2 - \omega'^2} d\omega' \tag{5.2}$$

where ω denotes the photon energy. This relation suggests that the knowledge of $f''(\omega)$ over a sufficiently wide energy region provides a method for evaluating the dispersion behavior of real part f'.

The energies of some characteristic x-rays such as Cu-Kα (9.048 keV) and Co-Kα (6.930 keV) are located near an absorption edge (8.332 keV) of Ni atom, and the anomalous dispersion factors become really sizable . For example, in the Ni atom f' = -2.96 and f" = -0.51 for the Cu-Kα radiation and the change of f' with respect to f° corresponds to 11 %. Thus the measurements of x-ray scattering intensity at energy near the absorption edge of the constituent atom gives an additional item of information about the atomic scale structure of desired materials. As shown in **Fig.5.2** the change in the x-ray anomalous dispersion factors of 3d transition metals and 4f rare earth metals is distinct for several characteristic Kα radiations. This implies that the anomalous x-ray scattering technique could be applicable to samples containing 3d transition metals or 4f rare earth metals using the characteristic Kα radiations produced by commercial x-ray targets.

As mentioned above, the anomalous x-ray scattering gives the energy dependent correction to the atomic scattering factor, thus we could change in weighting factor w_{ij} in eq.(4.4) required for separating the partial structure factors of multi-component disordered materials. We give here a brief description on the application of anomalous (resonance) x-ray scattering for determining the partial structure factors in a binary disordered system.

When the anomalous x-ray scattering occurs, the total scattering factor has the form of eq.(5.1) with the anomalous dispersion effect of f' and f". The square of the mean scattering factor $<f>^2$ and the mean square average of the scattering factor $<f^2>$ are then expressed by:

$$<f>^2 = [\Sigma c_i f_i (q,E)]^2 = <f><f^*> = <(f^o+f')>^2 + <f">^2 \tag{5.3}$$

$$<f^2> = \Sigma c_i f_i^2 (q,E) = <(f^o+f')^2> + <(f")^2> \tag{5.4}$$

The Laue monotonic scattering term in eq.(4.1) and the weighting factor w_{ij} in eq.(4.3) can also be written as follows:

$$[<f^2>-<f>^2] = c_1(1-c_1)f_1 f_1^* + c_2(1-c_2)f_2 f_2^*$$

$$- 2c_1 c_2 [(f_1^o+f_1')(f_2^o +f_2') + f_1" f_2"] \tag{5.5}$$

$$w_{11} = c_1^2 [(f_1^o+f_1')^2 + (f_1")^2]/<f>^2$$

$$w_{22} = c_2^2 [(f_2^o+f_2')^2 + (f_2")^2]/<f>^2 \tag{5.6}$$

$$w_{12} = c_1 c_2 [(f_1^o+f_1')(f_2^o+f_2') + f_1" f_2"]/<f>^2$$

The measurements of x-ray scattering intensity at two energies near the absorption edge of the constituent atom provide additional information of the so-called total structure factor of a binary dis-ordered system. (For example, the measurements with Cu-Kα and Co-Kα radiations in Fig.5.1). These data, when coupled with that obtained from the measurement at an energy away from the aborption edge, (for example, the measurement with Mo-Kα radiation in Fig.5.1) permits separation of the three partial structure factors in the manner of eq.(4.8).

In the anomalous x-ray scattering for most of the elements, the real part of the anomalous dispersion term f' is typically 20~30 % of the standard atomic scattering factor f^o at the K-shell absorption edge and f' appears to be substantially larger value (over 50 %) at the L-shell absorption edge, as exemplified by **Fig.5.3** of the Cs atom as an example. However, the energies of the characteristic Kα radia-tions are often not close enough to the absorption edges of the

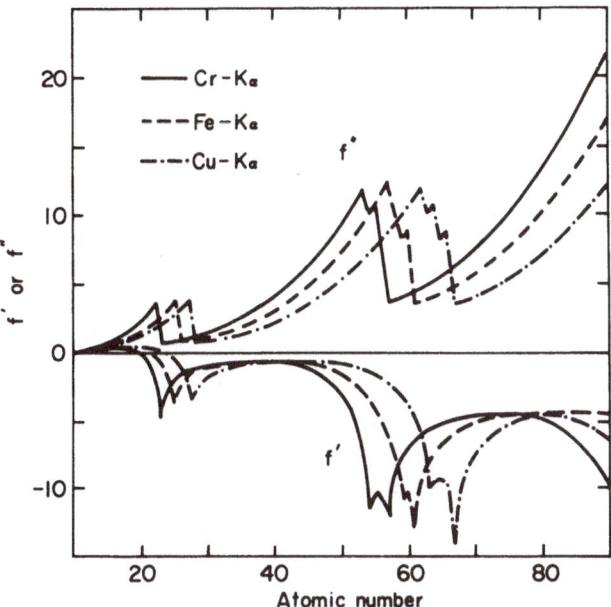

Fig.5.2 Anomalous dispersion factors of various elements for
characteristic Kα radiations of Cr, Fe and Cu taken
from the results of Cromer and Liberman (1970).

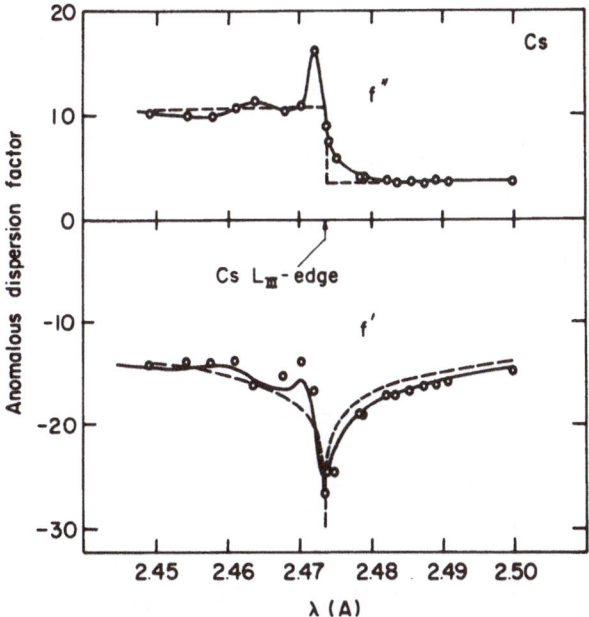

Fig.5.3 Anomalous dispersion factors of Cs atoms measured by
the single-crystal diffraction measurement (circles)
and calculated (broken lines) near the L_{III} edge. The
solid curve for f' is evaluated from f" data through
the dispersion relation (Templeton et al. 1980).

constituent atoms, so that the efficient use of the anomalous x-ray
scattering technique has been attained for only a relatively small
number of materials, as long as we use the Kα radiations produced by
commercial x-ray targets.

With respect to this subject, the energy dispersive measurements
appear to hold promise in reducing such difficulty by making available
a continuous energy spectrum and then enabling the use of an energy in
which the anomalous dispersion effect is the greatest. It is also
possible to carry out the particular experiment with two different
energies, for example λ_2 and λ_3 in Fig.5.1 where the imaginary part of
f" is almost constant and small, and only the change due to the real
part of f' is dominant or the combination of λ_1 and λ_3 in Fig.5.1
where only the imaginary component f" could be changed in the measurement, because of a constant value in the real component f'. The difference obtained in these two measurements contains information only
about atoms scattering x-rays anomalously. The angular scanning
measurement coupled with the energy dispersive mode (see **Appendix 2**)
using the energy sensitive solid state detector (often called as SSD)
gives the x-ray scattering intensity covering a wide energy range in
which the anomalous x-ray scattering is well detected. That is, three
dimensional spectrum $I(q,E)$ could be obtained by the angular scanning
measurement in the energy dispersive mode. As easily seen in eq.(5.1),
the $f^o(q)$ term depends only upon q. Therefore, the energy derivative
of measured x-ray scattering intensity $[\partial I(q,E)/\partial E]_q$ at a constant
value of q is dominated by the change arising from the energy
dependence of the anomalous dispersion factors, $(\partial f'/\partial E)_q$ and $(\partial f"/\partial E)_q$. Hence the fundamental relation for deriving the partial structure factors can be obtained in the following form, with respect to
the angular scanning measurements coupled with the energy dispersive
mode.

$$[\frac{\partial I'(q,E)}{\partial E}]_q \propto c_1^2[(f_1^o+f_1')\frac{\partial f_1'}{\partial E} + f"\frac{\partial f_1"}{\partial E}]a_{11}(q)$$

$$+ c_2^2[(f_2^o+f_2')\frac{\partial f_2'}{\partial E} + f_2"\frac{\partial f_2"}{\partial E}]a_{22}(q)$$

$$+ c_1c_2[(f_1^o+f_1')\frac{\partial f_2'}{\partial E} + f_1"\frac{\partial f_2"}{\partial E} + (f_2^o+f_2')\frac{\partial f_1'}{\partial E} + f_2"\frac{\partial f_1"}{\partial E}]a_{12}(q) \qquad (5.7)$$

where $I'(q,E)$ denotes the x-ray scattering intensity of structural
sensitive part of a binary disordered materials. The basic concept of

this energy derivative technique corresponds to the so-called **frequency modulated x-ray diffraction** first proposed by Shevchik (1977). When the energy of incident x-rays is tuned just close to an absorption edge for one of the constituent atoms, the energy derivative of the scattered intensity data is strongly affected by the interference from the corresponding constituent atom with a sensitivity greater than the case of the usual anomalous x-ray scattering measurement by the common characteristic $K\alpha$ radiations (Shevchik 1977, Munro 1982).

These interesting features clearly suggest that the use of a strong intensity x-ray source such as the synchrotron radiation could provide much valuable structural information for multi-component disordered materials by using the energy derivative mode with a combination of the anomalous x-ray scattering. This is also based on the following advantages:

(1) The very high intensity x-ray source of the synchrotron radiation with the well established and wide energy range of spectrum appears to enable us the use of an energy in which the anomalous dispersion effect is greatest. This contrasts to the characteristic $K\alpha$ radiations whose energies are often not close enough to the absorption edges of the desired elements, except for 3d transition metals and 4f rare earth metals.

(2) The synchrotron radiation provides not only high intensity but also clean white spectrum over a much wider energy range above 4 keV, compared with the conventional white x-ray source such as the W-target. Only the energy above 15 keV is generally used for measurement with a conventional W-target, because of the L-lines of the W atoms and $K\alpha$ and $K\beta$ lines of the Mo atoms arising from the impurities of filament. On the other hand, the synchrotron radiation is only limited by the absorption of the Be window ($\simeq$ 4 keV).

The usefulness of the anomalous x-ray scattering technique coupled with the synchrotron radiation source will be clearly demonstrated by the recent results on an amorphous $Mo_{50}Ni_{50}$ alloy (Aur et al 1983a) as given in the later section.

CHAPTER 6
THEORETICAL ASPECTS ON THE ANOMALOUS DISPERSION FACTORS OF X-RAYS

Various aspects on the anomalous x-ray scattering effects have been already discussed for several reasons (see for example, Ramaseshan and Abrahams 1975, Hosoya 1977, Gerward et al. 1979). The real component of the anomalous dispersion factor f' is required for the accurate structural analysis of both crystalline and non-crystalline systems, whereas the imaginary component f" represents the absorption and is necessary for the data processing for diffraction or EXAFS measurements. The complex correction factor f'+if" can also be used for the phase determination of the so-called crystal structure factor or the determination of the compositional short range order in multi-component disordered materials. In particular, the anomalous x-ray scattering with a high intensity x-ray source is likely to provide substantial progress in the structural study of disordered systems as shown in the previous chapter. Thus, in order for the anomalous x-ray scattering technique to reveal much valuable structural information, we must well characterize the magnitude and the energy dependence of the anomalous dispersion factors, f' and f", of x-rays.

The values of the anomalous dispersion factors have been theoretically calculated for six energies corresponding to the frequently used characteristic Kα radiations of Ag, Mo, Cu, Co, Fe and Cr (Dauben and Templeton 1955, Cromer 1965, Cromer and Liberman 1970). However, the energy variable measurement for the anomalous x-ray scattering technique can use a wide range of energy spectrum for structural determination, and this requires the knowledge of the anomalous dispersion factors as a continuous function of energy. The theoretical description for the anomalous dispersion factors of x-rays has been given in detail (see for example, James 1954, Fukamachi 1977), so that we give here a brief background on the theoretical aspects of these important dispersion factors as they are necessary for the efficient use of the anomalous x-ray scattering technique.

Following the discussion of James (1954) and Fukamachi (1977), the fundamental equations for evaluating the anomalous dispersion factors can be expressed as follows.

$$f'(\omega) = -\frac{1}{2} \int (\frac{dg_{oj}}{d\omega_{jo}}) \omega_{jo} \{\frac{\omega_{jo}-\omega}{(\omega_{jo}-\omega)^2+\gamma^2_{oj}/4} + \frac{\omega_{jo}+\omega}{(\omega_{jo}+\omega)^2+\gamma^2_{oj}/4}\} \, d\omega_{jo} \qquad (6.1)$$

$$f''(\omega) = \frac{1}{2} \int (\frac{dg_{oj}}{d\omega_{jo}}) \omega_{jo} \frac{\gamma_{oj}/2}{(\omega_{jo}-\omega)^2+\gamma^2_{oj}/4} \, d\omega_{jo} \qquad (6.2)$$

where ω corresponds to the photon energy and its subscript denotes the state of photon such as the initial (o) and the j-th scattering process. The γ_{oj} refers to the convoluted width of states o and j. The g_{oj} is the so-called oscillator strength which is defined by the following equation in the atomic unit ($h = m = e = 1$).

$$g_{oj} = (\frac{2}{\omega_{jo}}) \, |<\phi_o| \, \vec{e} \cdot \vec{P} \, | \, \phi_j> |^2 \qquad (6.3)$$

where ϕ_i is the wave function of atoms in the state of i, $\vec{e}$ is the unit vector and $\vec{P}$ is the momentum of electrons. When coupled with the simple approximation of $\gamma_{oj} \to 0$, eqs.(6.1) and (6.2) can easily be reduced to the following equations similar to the expression given by the classical theory (see for example, James 1954).

$$f'(\omega) = \int \frac{(dg_{oj}/d\omega_{jo}) \omega^2_{jo}}{\omega^2 - \omega^2_{jo}} \, d\omega_{jo} \qquad (6.4)$$

$$f''(\omega) = 2/\pi \int \omega_{jo} (\frac{dg_{oj}}{d\omega_{jo}}) \delta (\omega_{jo}-\omega) \, d\omega_{jo} = (\frac{\pi\omega}{2}) (\frac{dg}{d\omega}) \qquad (6.5)$$

Equation (6.5) gives the relation of $(dg_{oj}/d\omega_{jo}) = (2/\pi\omega_{jo}) \times [f''(\omega_{jo})]$ when $\gamma_{oj} \to 0$ and hence, the so-called **dispersion relation** is obtained.

$$f'(\omega) = -\frac{1}{\pi} \int [f''(\omega_{jo})]_{\gamma_{oj}\to 0} \{\frac{\omega_{jo}-\omega}{(\omega_{jo}-\omega)^2+\gamma^2_{jo}/4} + \frac{\omega_{jo}+\omega}{(\omega_{jo}+\omega)^2+\gamma^2_{jo}/4}\} d\omega_{jo} \qquad (6.6)$$

This can also be expressed by the following simpler form as already used in eq.(5.2).

$$f'(\omega) = \frac{2}{\pi} \int \frac{f''(\omega_{jo}) \omega_{jo}}{\omega^2 - \omega^2_{jo}} \, d\omega_{jo} \qquad (6.7)$$

The anomalous dispersion factors f' and f" can be estimated from the information on the energy dependence of the oscillator strength $(dg/d\omega)$ and in that regard, the following three methods may be mentioned for the major calculations of the anomalous dispersion factors that have been reported to date.

(1) **Non-relativistic quantum mechanics method** (Hönl 1933, see also James 1954).

(2) **Semi-empirical method** with the dispersion relation (Parrett and Hempstead 1954, Dauben and Templeton 1955).

(3) **Relativistic quantum mechanics method** (Cromer and Liberman 1970).

Honl (1933) proposed the method for evaluating the oscillator strength in terms of the non-relativistic wave function of hydrogen atom with some modification for the atoms containing more than two electrons and provided the following equation for the K-shell electrons.

$$\left(\frac{dg}{d\omega}\right)_K = \frac{2^8 e^{-4}}{9\omega_K} \left[\frac{4}{(1-\delta_K)^2}\left(\frac{\omega K}{\omega}\right)^3 - \frac{7}{(1-\delta_K)}\left(\frac{\omega K}{\omega}\right)^4\right] \tag{6.8}$$

where $\omega_K = \omega_0(1 - \delta_K)$, ω_0 is the threshold energy of absorption in the hydrogenic model and δ_K corresponds to the parameter containing the higher order electron effect on the ground state energy. Introducing eq.(6.8) into eqs.(6.4) and (6.5), one can estimate $f'(\omega)$ and $f''(\omega)$. This method has also been used for the L-shell electrons by Eisenlohr and Muller (1954) and to the M-shell electrons by Wagenfeld (1966) including quadrupole and higher order terms which are generally small (about 1 % of the dipole term). In the Hönl method, only transitions to positive energy final state are taken into account and a constant screening parameter is used for the inner K-shell electrons. Nevertheless, an agreement between the calculation and the experimental data has been obtained in some cases. However, it should be remembered that the results obtained by this method are no doubt less accurate than those of other methods, particularly for heavy elements, say Z > 30 where Z is the atomic number, because of the use of non-relativistic wave functions.

Even in the framework of classical mechanics, it is possible to calculate the anomalous dispersion factors $f'(\omega)$ and $f''(\omega)$, when coupled with the experimental absorption data, because the oscillator strength can be connected with the energy dependence of the linear absorption coefficient. This is readily explained by the relationship between the macroscopic refractive index and the linear absorption coefficient of x-rays (see for example, Miyake 1969). Along the line of this concept, Parratt and Hempstead (1954) proposed the simple but useful semi-empirical method for calculating $f'(\omega)$ and $f''(\omega)$. The measured absorption data at an energy ω is known to be described by the following simple form.

$$\left. \begin{array}{ll} \mu(\omega) = \mu_K (\omega_K/\omega)^{n_K} & \omega > \omega_K \\[2em] \quad\;\; = 0 & \omega < \omega_K \end{array} \right\} \qquad (6.9)$$

where ω_K is the energy of the absorption edge, μ_K is the linear absorption coefficient at an energy of ω_K and n_K corresponds to the polynominal index. The parameters μ_K and n_K can be determined by fitting a curve of eq.(6.9) to measured absorption data. Since the energy derivative of the oscillator strength $(dg/d\omega)$ is proportional to the linear absorption coefficient term $\mu(\omega)$, the following form, similar to eq.(6.9), may also be well recognized.

$$(dg/d\omega)_K = g'_K (\omega_K/\omega)^{p_K} \qquad\qquad \omega > \omega_K \qquad (6.10)$$

When the quantity g_K is defined by the following equation,

$$g_K = \int_{\omega_K}^{\infty} (dg/d\omega)\; d\omega \qquad\qquad (6.11)$$

we obtain,

$$g'_K = (p_K - 1)\; \frac{g_K}{\omega_K} \qquad\qquad (6.12)$$

The energy derivative of the oscillator strength can then be given by,

$$\left(\frac{dg}{d\omega}\right)_K = \frac{p_K - 1}{\omega_K}\; g_K \left(\frac{\omega_K}{\omega}\right)^{p_K} = \frac{c^o}{2\pi^2}\mu_K \left(\frac{\omega_K}{\omega}\right)^{p_K} \qquad \omega > \omega_K \qquad (6.13)$$

where c^o is the speed of light.

On the basis of this idea, Parratt and Hempstead (1954) subtracted the fitting part to the experimental data (g_K or μ_K) an numerically integrated the residue by a similar manner of eqs.(6.4) and (6.5) and then gave the equations for evaluating the anomalous dispersion factors with an appropriate value of p_K. Following their procedure, Dauben and Templeton (1955) calculated the anomalous dispersion factors for three common characteristic radiations (Cr-Kα, Cu-Kα, and Mo-Kα) by using the measured linear absorption coefficient $\mu(\omega)$, and the results have been compiled in the International Tables for X-ray Crystallography. The values reported by Dauben and Templeton (1955) include the angular variation of f' and f" by multiplying the dispersion correction of each shell. However, it should be remembered that this angular dependence has no physical significance at least theoretically, because the semi-empirical method accounts only for the dipole approximation which gives no angular dependence (see for example, Wagenfeld 1975).

Cromer (1965) also evaluated the anomalous dispersion factors for several characteristic radiations along the line proposed by Parratt and Hempstead (1954). In his case, the oscillator strength g_K in eq.(6.13) was theoretically calculated from the wave function of the Hartree type with the Thomas-Reich-Kuhn sum rule. The physical significance of this semi-empirical method depends mainly upon the accuracy of the parameter p_K (or n_K) for reproducing the linear absorption coefficient obtained experimentally.

Parratt and Hempstead (1954) proposed the following values; $p_{1s} = 11/4 = 2.75$, $p_{2s} = 7/3 = 2.33$ and $p_{others} = 5/2 = 2.5$. However, the assumption that the K-shell electrons possess the same value of 2.75 for all cases is no doubt a crude approximation. It may also be worth mentioning from the results of Weiss (1966) that the change in $p_K = 2.7 \pm 0.1$ produces the variation of the order of $\pm$ 10 % in the resulting f'(ω).

An attempt by Cromer and Liberman (1970) has been made to theoretically evaluate the anomalous dispersion factors for most of the elements (Z = 3 - 98) by using the relativistic wave function. Their expressions are of considerable length and somewhat complicated, but the basic concept of their method refers to the theoretical evaluation of the oscillator strength by using the more realistic wave function of atoms in the respective state (see eqs.(6.3)). The anomalous dispersion factors f' and f" are then calculated in the same manner as the other two methods employing eqs.(6.4) and (6.5). However, in contrast to the semi-empirical method, the integral for these equations has been evaluated numerically without approximation to the form of the cross section vs energy curve expressed by eq. (6.9) or

(6.10). Cromer and Liberman (1970) used the relativistic wave function of the Dirac-Slater type and the procedure for evaluating the relativistic photoelectron absorption coefficient given by Brysk and Zerby (1968) and provided the anomalous dispersion factors of many elements for five characteristic Kα radiations. Their results for both f' and f" show a reasonable agreement with the experimental data as shown in **Table 6.1.**

The values of the anomalous dispersion factors given by this Cromer-Liberman's scheme are probably the best knowledge in the presently available information, because of the rigorous and careful treatment for the evaluation process of f' and f". For convenience of future investigations using the anomalous x-ray scattering (AXS) and

Table 6.1 Comparison of experimental and calculated anomalous dispersion factors for some elements (Cromer and Liberman(CL), 1970; Gerward et al.(G), 1979).

	Radiation	Element	Calculation	Experiment
f'	Cu-Kα Mo-Kα Ag-Kα	Ge (CL)	−1.803 0.082 0.142	−1.79 0.08 0.27
f"	Cu-Kα Mo-Kα Ag-Kα	Ge (CL)	0.0 1.588 1.054	0.0 1.58 1.06
f'	Cu-Kα	Si P K	0.330 0.434 1.066	0.31 0.44 1.04
f"	Mo-Kα	Si P K (CL)	 0.095 0.250	 0.107 0.271
f'	Cr-Kα Fe-Kα Co-Kα Cu-Kα Mo-Kα Ag-Kα	Si (G)	0.381 0.337 0.313 0.270 0.098 0.068	0.389 0.344 0.320 0.274 0.099 0.070
f"	Cr-Kα Fe-Kα Co-Kα Cu-Kα Mo-Kα Ag-Kα	Si (G)	0.693 0.509 0.438 0.330 0.071 0.043	0.696 0.510 0.440 0.331 0.071 0.044

the energy dispersive x-ray diffraction (EDXD) technique, the energies of absorption edges in keV unit for various elements are summarized in **Appendix 4.** The energy dependence of the anomalous dispersion factors calculated by the Cromer-Liberman's scheme are also given in Appendix 4 for various elements with the energy region between 1 keV and 50 keV, because such information are not covered in the previous specialized monographs or common databook for x-ray crystallography.

All three theoretical calculations regarding the anomalous dispersion factors f' and f" are restricted to only free (isolated) atoms at the present time and hence none of them include the near-edge

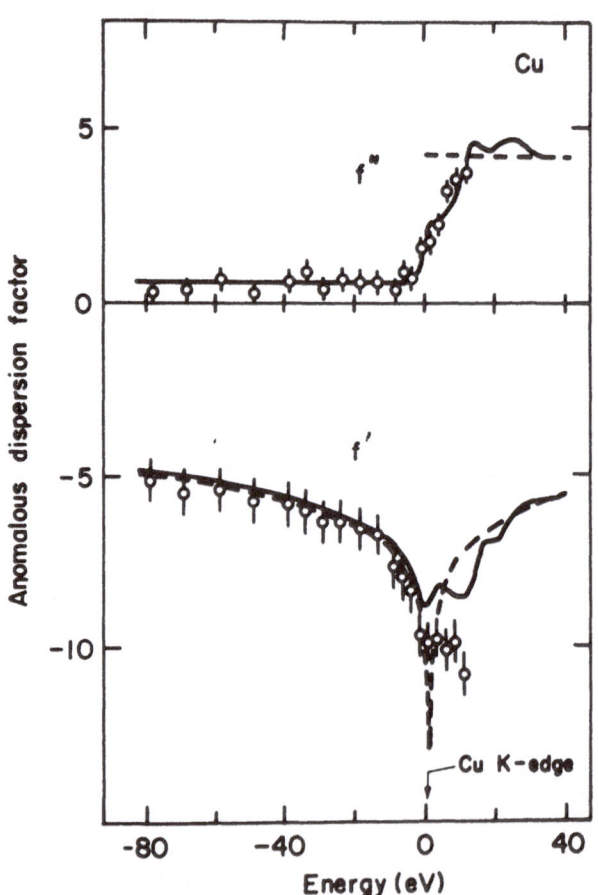

Fig.6.1 Anomalous dispersion factors of Cu atoms measured by the reflectivity measurement (circles) and caclulated (broken lines) near the K edge. The solid curve for f' is evaluated from f" data through the dispersion relation (Fukamachi et al. 1978).

phenomena such as the white line, the edge shift and the so-called EXAFS, frequently observed at the threshold of the absorption edge. Such characteristic fine structures appear significantly in the narrow energy region, particularly in the higher energy side of the absorption edge, as exemplified by **Fig.6.1** using the crystalline Cu foil as an example (Fukamachi et al 1978). In this regard, the values calculated by the above three methods do not predict any fine structure in $f'(\omega)$ and $f''(\omega)$ associated with the near-edge phenomena. For this purpose, the alternative rigorous consideration such as the band structure effect and the chemical environmental correlation around atoms scattering x-ray anomalously should be taken into account for the wave functions required for evaluating the oscillator strength (see eq.(6.3)). In the near-edge region, the wave functions are known to strongly vary from one state to the other (see for example, James 1954). Thus further calculation for the anomalous dispersion factors, particularly in the energy region just beyond the absorption edge, is required to predict the near-edge phenomena. This should be one of the most important future research subjects in the anomalous x-ray scattering or in the EXAFS measurement. However, much further experimental works is highly desirable, because a sufficient amount of information about near-edge phenomena has not been explored yet.

CHAPTER 7

EXPERIMENTAL DETERMINATION OF THE ANOMALOUS DISPERSION FACTORS

As shown in the previous chapter, the anomalous dispersion factors f' and f" calculated by the Cromer-Liberman scheme (1970) or the semi-empirical scheme of Cromer (1965) with the relativistic wave function agree well with the experimental data determined for various elements mostly by using the characteristic Kα radiations. Such agreement has also been found in a wider energy region except for the particular near-edge region as exemplified by **Fig.7.1** using the results of a Zn atom in hemimorphite (Hosoya 1979). With these facts in mind, the theoretical values of f' and f" evaluated for an isolated atom (see for example, Cromer and Liberman 1970) permit the use of data processing on the structural analysis of various materials, except for the particular energy narrow energy region such as the energy just abobe the absorption edge by about 50 - 100 eV. Namely the so-called **chemical transferability** may be well accepted in the energy region where the physical significance of the near-edge phenomena is not severe. The lower energy side of the absorption edge is likely to fall under this category as easily seen from an example of Fig.7.1.

On the other hand, the use of the energy dispersive x-ray diffraction technique with a continuous spectrum (see also **Appendix 2**) enables us to treat an energy region where the near-edge phenomena is appreciable. The synchrotron radiation as a light source also provides not only an extremely high intensity but also a easily tunable and highly monochromatic x-ray source. These major technical advances obtained only recently appear to perform the experiment at an energy in which the anomalous x-ray scattering is the greatest. This also inevitably emphasizes an increasing need for making measurement of the anomalous dispersion factors for the respective materials of interest, because one can no longer rely upon the theoretical values of f' and f" calculated for an isolated atom in the narrow energy region where the near-edge phenomena is appreciable. The experimental verification and its understanding of the near edge phenomena, are still far from complete. This includes the detailed information for f'(E) and f"(E) in the higher energy side (by about 50 - 100 eV) of the absorption edge.

Various method have been attempted to determine the anomalous dispersion factors of x-rays and the respective technique, of course, has its own advantages and disadvantages. Although there is no result which gives any definite conclusion regarding the superiority of one method to the others at the present time, some methods indicate good feasibility, in parallel with recent progress of the energy dispersive technique with a synchrotron radiation source. Thus this research field is expected to grow very remarkably in the next five to ten years. We give here a brief background of the selected examples for experimentally determining the anomalous dispersion factors, for future application of the anomalous x-ray scattering to the structural study of various materials. The methods for the measurement of f' and f" used in the literature can be classified into one of the following two types, although there are differences in detail. One is the **OPTICAL METHOD** and the other is the **INTENSITY MEASUREMENT METHOD.**

One of the typical optical methods is the measurement for the deviation from unity in the refractive index in a manner similar to the visible (light) region (see for example, James 1954). Another typical optical method is the use of the x-ray interferometry (see for example, Bonse and Materlik 1976). Both techniques determine the real

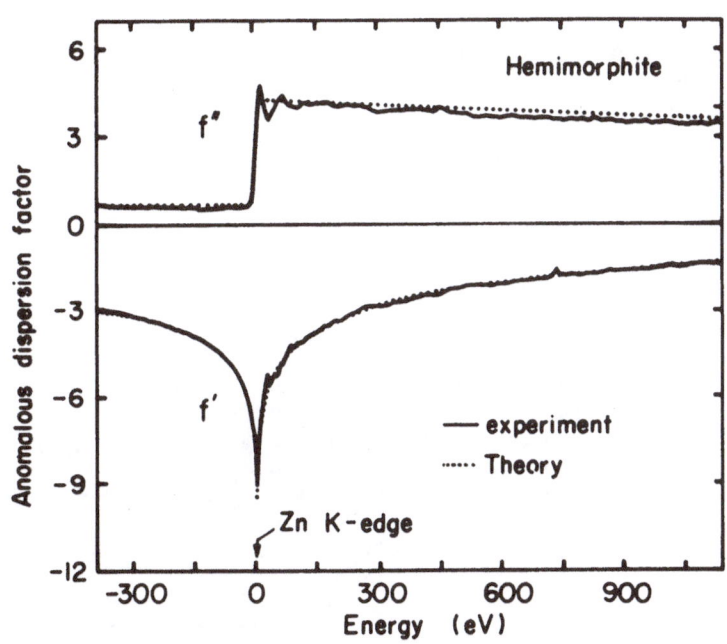

Fig.7.1 Energy dependence of anomalous dispersion factors for
Zn atoms in hemimorphite near the K edge. The values
of f' are evaluated from the absorption data of f"
through the dispersion relation (Hosoya 1979).

component of the anomalous dispersion factors f' as a phase shift, because the absorption of sample materials related to the imaginary component f" gives only the reduction of x-ray intensity in these measurements. For example, the absorption broadens the cutoff in the total reflection or gives no change in the position of the interference patterns. Recently Bonse and his coworkers (1976) constructed the x-ray interferomater device as schematically shown in **Fig.7.2** and revealed the very detailed near-edge phenomena in crystalline Ni or Cu foil, with the help of a synchrotron radiation source at DESY (Germany). Their results of the crystalline Cu foil are given in **Fig.7.3** (Bonse et al. 1982). The absorption curve related to the f" component was also measured by placing a Cu foil in front of the detector with the phase shifter plane removed from the interfering x-ray beam. The four maxima are found in the absorption curve just above the edge. This fine structure has frequently been referred to as **XANES** (X-ray Absorption Near Edge Structure) which principally differs from the well-known EXAFS (see for example, Bonse et al 1982). Such fine structural change contrasts with the information in the lower energy

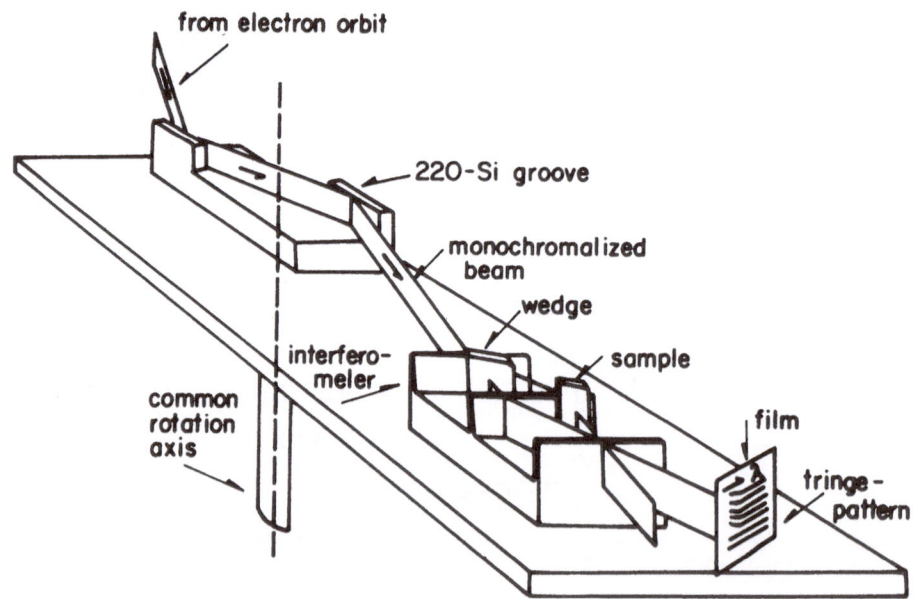

Fig.7.2 Schematic diagram of x-ray interferometer (Bonse and Materlik 1976).

side of the absorption edge where only the monotonic energy dependence is detected. A similar energy dependence has also been observed in the reflectivity measurement by Fukamachi et al. (1978) as shown in Fig.6.1, although a direct comparison is not allowed due to the large difference in energy resolution and experimental uncertainty. Further experiments for various materials should give successful and important information regarding the near-edge phenomena in the near future.

On the other hand, most of the intensity measurement methods such as the diffracted intensity measurement for the Friedel pair reflections (see for example, Zachariasen 1965) and the absorption measure-

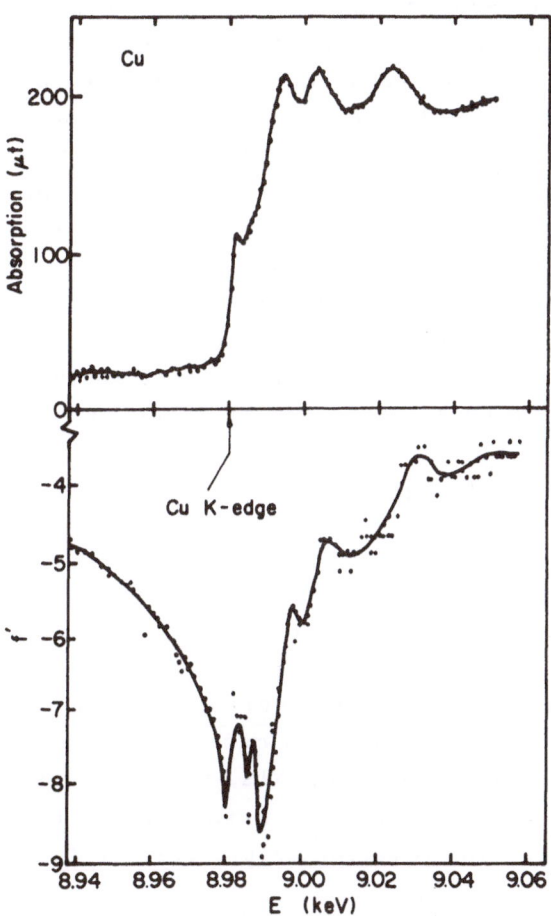

Fig.7.3 Measurement of anomalous dispersion factor f' (bottom) and absorption μt (top) as a function of energy for Cu foil near the K edge (Bonse et al. 1982).

ments by the two crystals or one crystal with solid state detector system (Hosoya and Yamagishi 1966, Sirota 1969) are always affected by the absorption of sample materials and thus the data processing for extracting f' and f" from measured intensity data seems to be somewhat complicated, compared with the case by the optical method. However, when the absorption measurement could be carried out with a reasonably higher degree of accuracy, a relatively simple method is available for determining both anomalous dispersion components f' and f" by applying the so-called **dispersion relation**.

The theoretical basis of this indirect method, has been discussed in detail by Kawamura and Fukamachi (1978) and its usefulness has now been conceptually well-recognized (see for example, Fuoss et al 1980, Bonse et al. 1982). The experimental uncertainty in the absorption experiment of x-rays has been extremely reduced recently in parallel with the significant technical progress in both detecting systems and x-ray sources such as the solid state detector and a synchrotron radiation. Therefore, this indirect method appears to be very convenient for determining the anomalous dispersion factors of the particular sample being investigated. It may be helpful to recall the essential points of this indirect method.

The imaginary component of the anomalous scattering of x-rays $f"(\omega)$ at an energy of ω is directly related to the linear absorption coefficient $\mu(\omega)$ through the following equation.

$$f"(\omega) = \frac{m\, c^O M}{4\pi N^2 e\, \rho} \mu(\omega) \qquad (7.1)$$

where m and e are the electron mass and charge, respectively, M is the atomic weight, N is the Avogadro's number, ρ is the density of the materials and c^O is the speed of light. It may also be noted that the f" curve as a function of energy corresponds to the so-called EXAFS curve which has recently received much attention for determining the fine structure of various materials. **In this regard, the EXAFS is a part of the anomalous x-ray scattering.** The real component of $f'(\omega)$ is given by the dispersion relation as proposed by Kawamura and Fukamachi (1978).

$$f'(\omega) = \frac{2}{\pi} \int_0^\infty \frac{f"(\omega)}{\omega^2 - \omega'^2}\, d\omega \qquad (7.2)$$

Figure 7.4 gives the anomalous dispersion factors of a Ga atom in the

crystalline GaAs determined by this combination method of the absorption measurement for the Friedel pair reflections (Fukamachi et al.
1979).

It is readily seen from eq.(7.2) that the experimental uncertainty of the real component f'(ω) is directly attributed to that of the
imaginary component f"(ω) and thus an accurate measurement for the
absorption of x-rays should be made for this combination method. The
homogeneity of a sample and the energy resolution of a monochromatic
x-ray source are known to give the significant contribution to the
accuracy of the absorption measurement, so that the experimental set-
up should be chosen to minimize these factors. The energy dependence
of the f"(ω) curve should be also measured in a wide energy region for

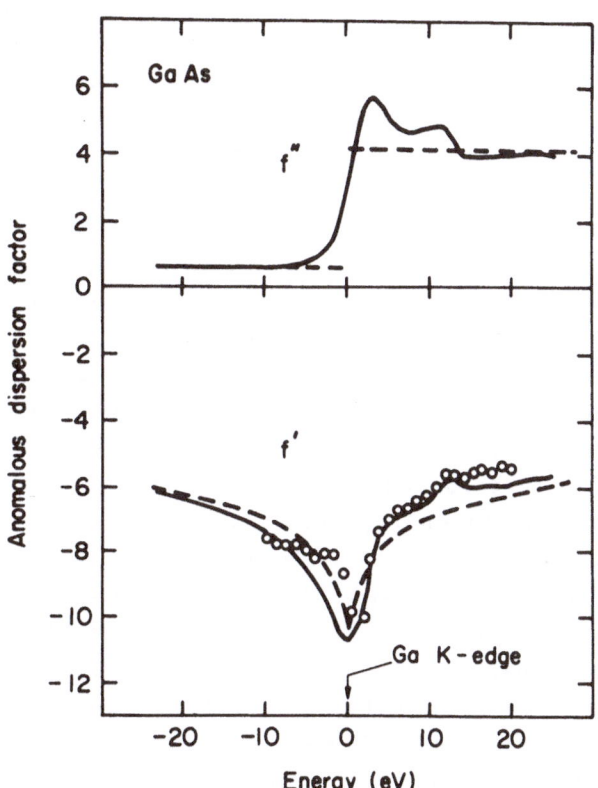

Fig.7.4 Energy dependence of anomalous dispersion factors f'
and f" for Ga atoms in GaAs near the K edge. Intensity
measurement for the Friedel pairs (open circles),
theory by the Hönl method (broken lines), The solid
curve for f' is evaluated from f" data through the
dispersion relation (Fukamachi et al. 1979).

the integration of eq.(7.2), because the near-edge values of $f'(\omega)$ are known to depend mainly upon the form of the $f''(\omega)$ curve in the energy region where the drastic change appears. The integration limits of eq.(7.2) give an influence on the resultant magnitude of the f' values. However, the fine structure in $f'(\omega)$ does not change with the variation of the integration limits (see for example, Kawamura and Fukamachi 1978). The following points may also be worth of note. The measured absorption data are sometimes affected by the attenuation arising from some coherent scattering intensity of a sample, such as the small angle scattering related to an extinction of the transmitted x-ray beam as found in the measurement of graphite (see for example, Chipman 1955). However, the small angle scattering intensity is known to be generally small in disordered materials.

The accurate determination of the absolute values for the anomalous dispersion factors f' and f'' is one of the difficult experiments. On the other hand, the relative value determination, such as the energy dependence of $f'(\omega)$ and $f''(\omega)$ may be easier to carry out for both crystalline and non-crystalline systems. The structure factor (or interference function) of disordered materials is well known to approach unity in the large wave vector (q) region at any energy (see chapters 2 and 3). Thus, the variations in the structure factor beyond $q = 12$ A^{-1} are generally enough to determine the anomalous dispersion factors assuming that the structure factor is equal to unity. With the help of this particular feature, the energy dependence of the anomalous dispersion factors for disordered materials could be estimated by superimposing the relative intensity measurement at several different energies near the absorption edge of a specific element in the sample on the reference value at energy away from the absorption edge. Such reference energy is easily chosen from the data given in **Appendix 4.**

The experimental determination of the anomalous dispersion factors of x-rays is not established yet and thus the experimental values of f' and f'' may be less accurate in some cases than that required for the anomalous x-ray scattering experiments (see for example, Fuoss et al. 1980). However, it may be concluded for structural analysis of disordered materials that the combination method of the absorption measurement and the dispersion relation is one way to determine the anomalous dispersion factors of a desired sample, at least in cases when other techniques such as the optical method is found to be technically difficult or before the full potential of any technique presently proposed can be assessed as a reliable tool for determining the anomalous dispersion factors.

CHAPTER 8

SELECTED EXAMPLES OF STRUCTURAL DETERMINATION USING ANOMALOUS (RESONANCE) X-RAY SCATTERING

The anomalous x-ray scattering has been known for a long time and a large amount of both experimental and theoretical effort has been devoted to the understanding of the anomalous (resonance) scattering phenomena and its application to the structural determination of various materials. In the last ten years, significant progress has been made, although the quantitative information is often accepted with some reservation, mainly due to the experimental difficulties. Nevertheless, the anomalous x-ray scattering has recently drawn a great deal of attention, because the use of the synchrotron radiation as an x-ray source in this technique is likely to reduce the experimental difficulties un-solved by a conventional x-ray source. Thus the anomalous x-ray scattering should become, in the near future, one of the most reliable and powerful tools for determining the fine structure of various materials and its potential application to the structural study of disordered materials is now well-recognized.

The main purpose of this chapter is to present some selected recent experimental results on the structural characterization by the **anomalous x-ray scattering** technique including present problems and future directions on the novel application of this technique for structural study of disordered materials. These topics include not only liquids and amorphous solids but also solid fast ion cunductors because their ionic transport properties imply that the cations in some solid fast ion conductors are in the liquid-like disordered state, although they are macroscopically still in the typical solid state. As shown later, the anomalous x-ray scattering provides a straightforward answer on such questions.

8.1 Crystalline Materials

For convenience of the fuller understanding of the basic idea and usefulness of the anomalous x-ray scattering, we give first a few examples of the structural determination of crystalline materials by applying this novel technique of the anomalous (resonance) x-ray scattering.

(A) Noncentrosymmetric Single Crystals

One of the typical application of the anomalous x-ray scattering to structural study of various materials is known to determine the crystallographic polarity of noncentrosymmetric systems such as ZnO and GaP single crystals. The usual crystallographic structure factor $F(\vec{h})$ is given by the following equation.

$$F(\vec{h}) = \sum_j f_j \exp(2\pi i \vec{h} \cdot \vec{r}_j) = \sum_j (f^0 + f' + if'')_j \exp(hx_j + ky_j + lz_j) \qquad (8.1)$$

where $\vec{h}$ corresponds to the reciprocal lattice points in the wave vector $(\vec{q})$, characterized by the so-called Miller index (h,k,l) in the crystal and f_j and $\vec{r}_j$ are the usual atomic scattering factor and the spatial coordinates (x_j, y_j, z_j) in the unit cell for the j-atom, respectively.

When the anomalous x-ray scattering of the constituent atom becomes significant in a noncentrosymmetric single crystal, the so-called Friedel's rule of $F^*F(h,k,l) = F^*F(\bar{h},\bar{k},\bar{l})$ could be no longer conserved for all diffraction planes. Such a relation is readily seen from the schematic diagram of the difference in the crystallographic structure factors of the opposite sides of a certain diffraction plane in **Fig.8.1**. However, the phase of the scattered x-rays is not directly detected and thus the effect of the imaginary component of the anomalous dispersion can be made apparent only through the intensity difference caused by the interference with a real component. A striking evidence of such interference effect is exemplified by the marterials with a wurtzite-type structure, for example, ZnO and CdS.

Fora wurtzite-type configuration of atoms, the following relations are obtained with respect to the crystallrographic structure factor for the {001} family of the reflections.

$$
\left.
\begin{array}{lll}
F^*F(001) = F^*F(00\bar{1}) = 0 & \quad 1 = \text{odd number} \\[6pt]
F^*F(001) = F^*F(00\bar{1}) & \quad 1 = 4n \\[6pt]
F^*F(001) \neq F^*F(00\bar{1}) & \quad \pm 1 = 4n-2
\end{array}
\right\} \qquad (8.2)
$$

where n is zero or an integer. Equation (8.2) indicates that the (004) and (008) pairs should give equal intensities whereas the (002) and

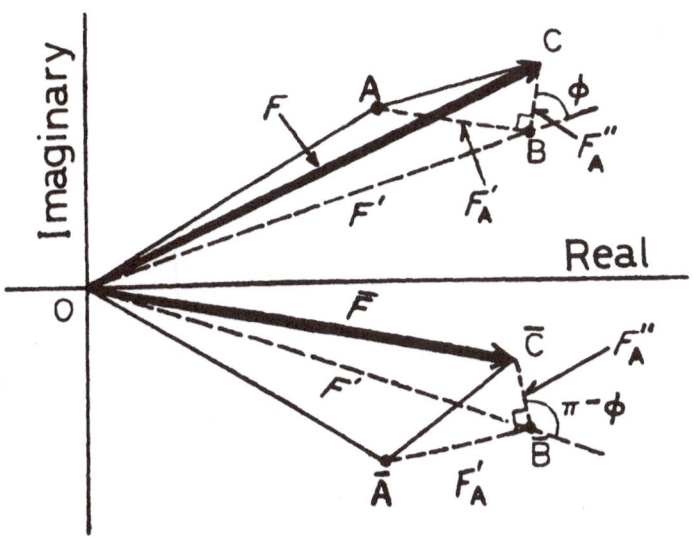

Fig.8.1 Schematic diagram of the difference in crystallographic
structure factor of the oposite sides of a certain
diffraction plane arising from the anomalous dispersion
effect.

(006) pairs give a **nonequivalence** of diffracted intensities, $I_{(h,k,l)}$ $= I_{(\bar{h},\bar{k},\bar{l})}$, where the constituent atoms scatter x-rays anomalously.

The first application of this technique was reported by Nishikawa and Matsukawa in 1928 regarding the determination of the crystallographic polarity of a ZnS single crystal using the K-absorption edge of Zn atoms, and a similar analysis has been carried out for structural characterization of various materials up to the present. Some marked examples may be as follows.

Mariano and Hanneman (1963) measured the continuous spectrum across the K-absorption edge (λ_K = 1.283 A) of Zn atoms produced by the (002) and (00$\bar{2}$) planes of a ZnO single crystal using a white x-ray source. As shown in **Fig.8.2**, a failure of the Friedel's rule $I_{(002)}$ = $I_{(00\bar{2})}$ is obvious in the energy region (λ = 1.263 A) where f_{Zn} shows a significant value (about 3.4). On the other hand, such deviation from the Friedel's rule is not detected in the energy region (λ = 1.303 A) corresponding to the lower energy side of the K-absorption edge of Zn atoms where f_{Zn} is small and nearly constant (see also Fig.7.1).

In this regard, Hosoya and Fukamachi (1973) proposed a new technique of the energy dispersive mode coupled with a white x-ray source

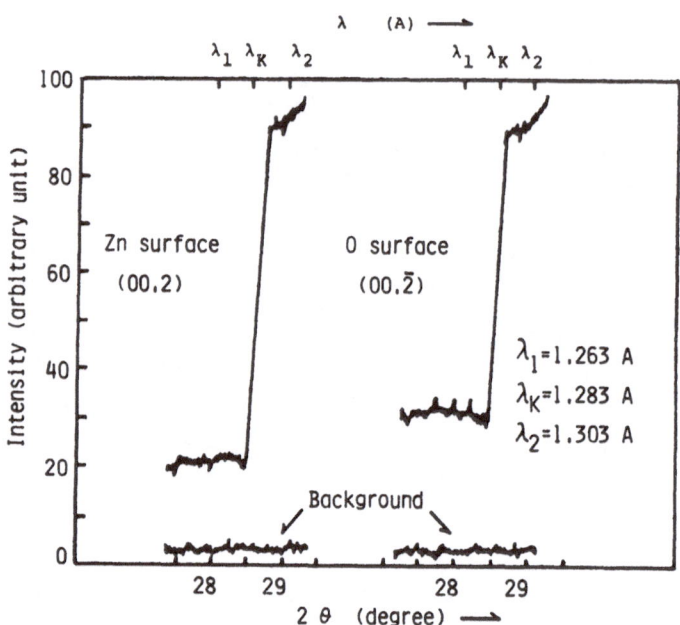

Fig.8.2 Continuous x-ray scattering spectrum across the K edge
of Zn atoms in ZnO single crystal (Mariano and Henneman
1963).

and a solid state detector and the validity of their technique has
been elegantly demonstrated by the measurement for determining the
crystallographic polarity of a GaP single crystal. The use of the
energy dispersive mode enables one to easily tune the relevant energy
in which the anomalous dispersion effect is appreciable for the particular diffraction plane by changing the scattering angle of the
spectrometer, as well as the effective elimination of the fluorescent
radiation. It should be remembered that the fluorescent radiation
contribution to measured intensity is occasionally siginificant in the
experiment with energies near the absorption edge. This point will be
described again in the latter section of 8.3. **Figure 8.3** shows the
results of a GaP single crystal in which the diffraction peaks arising
from {333} planes correspond to the energy region close to the K-
absorption edge of Ga atoms (Hosoya and Fukamachi 1973). A pair of
(333) and ($\bar{3}\bar{3}\bar{3}$) reflections indicate markedly different intensities
mainly due to the anomalous scattering of Ga atoms.

On the basis of these experimental results, one can discuss the
etching behavior, crystal morphology in opposite polar directions and

others for the noncentrosymmetric materials.

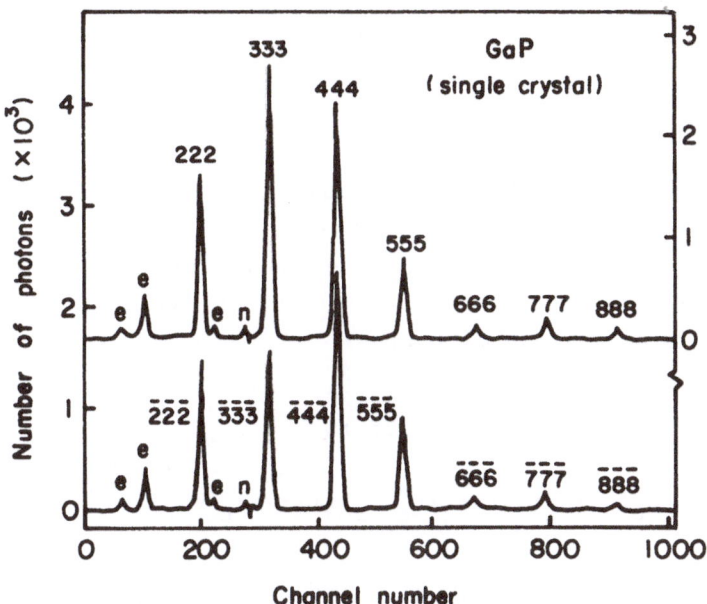

Fig.8.3 Comparison of (hhh) with ($\bar{h}\bar{h}\bar{h}$) reflections of GaP single
crystal. A pair of (333) and ($\bar{3}\bar{3}\bar{3}$) reflections shows
markedly different intensity due to the anomalous dispersion
effect of Ga atoms (Hosoya and Fukamachi 1973).

(B) Ordered Polycrystalline Alloys

The good utility of the anomalous x-ray scattering has also been
confirmed in structural investigation of polycrystalline samples,
particularly so-called ordered alloys and oxides with spinel-type
structure. Even in the case where the atomic number of the sample
constituents is similar, the variation in the crystallographic structure factor can be made, as long as the energy of the incident x-rays
is chosen to be close enough to the absorption edges of the constituents. It may be helpful to recall some essential points for these
structural investigation as given below.

Let us consider a binary polycrystalline alloy containing A and B
atoms with the composition c_A and c_B, respectively. When there exists
two sublattice denoted by the suffix 1 and 2 in this crystalline alloy

and then the number of sites in each sublattice is given by N_1 and N_2, the following relations can be obtained.

$$N_1^A + N_2^A = Nc_A, \qquad N_2^B + N_2^B = Nc_B$$

$$c_A + c_B = 1, \qquad N_A + N_B = N \qquad \left.\vphantom{\begin{array}{c}1\\1\end{array}}\right\} \qquad (8.3)$$

where N_j^α is the number of sites in sublattice j occupied by the α-atom and N corresponds to the total number of sites in an alloy. By using the expression of $P_j^\alpha = N_j^\alpha/N_j$ and the atomic scattering factor f_A and f_B, the so-called crystallographic structure factor of this crystal-line alloy may be written as follows.

$$F = (P_1^A f_A + P_1^A f_B)S_1 + (P_2^A f_A + P_2^B f_B)S_2 \qquad (8.4)$$

$$= (c_A f_A + c_B f_B)(S_1+S_2) + (f_A-f_B)\{(1-v)S_1-vS_2\}\frac{(P_1^A-c_A)}{(1-v)} \qquad (8.5)$$

where S_1 and S_2 are the geometrical factor of sublattice 1 and 2, respectively and $v = N_1/N$. The following points may be noteworthy. The first term in eq.(8.5) corresponds to the contribution for the case where the A and B atom distribution in the sublattice is perfectly random whereas the second term in eq.(8.5) can be attributed to the contribution of the perfect ordered case such as the superlattice.

In addition, the term of $(P_1^A - c_A)/(1 - v)$ is frequently referred to as the **degree of long-range order** (see for example, Warren 1969, Klug and Alexander 1974). It is easily seen from eq.(8.5) that the deviation from the limiting case such as the perfectly random struc-ture (or perfectly ordered structure) should be proportional to the difference of $|f_A - f_B|$ in the atomic scattering factors of the sample constituents. Thus, the use of the incident x-ray energy where the anomalous scattering is significant appears to provide the intensity difference for the particular diffraction planes by making available the variation of the crystallographic structure factor F, and then enabling us the structural characterization of sample materials.

By this method with the help of the anomalous x-ray scattering, the superlattice structure of various materials has been determined and some typical examples are summarized in **Table 8.1** together with the used characteristic Kα radiations.

Table 8.1 Examples for structural determination of polycrystalline
materials by the anomalous x-ray scattering.

Materials	Radiations	References
Cu_2MnAl	Fe-Kα, Cu-Kα, Zn-Kα	a
Ni_3Fe	Co-Kα	b
FeCo	Co-Kα	c
Ni_2Cr	Cr-Kα	d
Cobalt ferrite	Co-Kα	e
Nickel ferrite	Fe-Kα, Fe-Kα	f

(a) A.J.Bradley and J.W.Rodgers: Proc. Roy. Soc. A144, 340(1934).

(b) R.J.Wakelin and E.L.Yates: Proc. Phys. Soc. B66, 221(1953).

(c) W.E.Ellis and E.S.Greiner: Trans. ASM, 29, 425(1941).

(d) H.G.Baer: Z.Metallkde. 49, 614(1958).

(e) F.Bertaut: Compt. Rend. 231, 88(1950).

(f) L.P.Skolnick, S.Kondo and L.R.Lavine: J.Appl. Phys. 29, 198(1958).

(C) Cation Distribution in Ferrites

The cation distribution in many ferrites of the so-called spinel-
type structure expressed by $MO \cdot Fe_2O_3$ is generally difficult to determine by the usual x-ray diffraction technique, because the scattering
ability of the components (M) such as Mn, Co and Ni is close to that
of host element of Fe. However, an appreciable difference in their
cyrstallographic structure factor can be made by the anomalous dispersion effect. The essential procedure for determining the cation
distribution in ferrites is almost identical to those described in the
previous sections. Hence, only some examples analyzed by this technique are summarized in **Table 8.1**. A similar approach using the Fe-Kα
and Cr-Kα radiations has been attempted to determine the position of
a Cr atom in the cementite-type structure of $Fe_2(Fe,Cr)C$ by Kudiekla
and Moller (1963).

There exists a number of structural studies on crystalline materials and macromolecular complexes by the anomalous x-ray scattering.

However, it is not the author's intension here to cover all these topics in detail. With respect to such information, the reader may wish to refer to other specialized monographs or review articles (see for example, Ramaseshan 1964, Srinivasan 1972, Ramaseshan and Abrahams 1975, Hosoya 1977, Blasie and Stamatoff 1981) in which the experimental results on the structure of disordered materials have not been covered.

8.2 Non-Crystalline Materials

Many interesting disordered materials contain more than two kinds of atoms. Therefore, one of the most important research subjects in this field is known to evaluate the partial RDFs which correspond to the Fourier transform of the partial structure factors (or partial interference functions) for the individual chemical constituents. The use of the anomalous x-ray scattering is one way with which to permit the separation of partial structure factors from measured intensity data. Thus the presently available information is given below with respect to the structural investigation of disordered materials by using the anomalous dispersion effect.

The principle and its usefulness of the anomalous x-ray scattering for determining the fine structure of disordered materials have been suggested several times (see for example, Krogh-Moe 1966, Groubert and Regis 1967, Ramesh and Ramaseshan 1971), but successful results have been obtained only recently, in parallel with current technical improvements in a x-ray source and the detecting devices of x-ray photons. The fundamental equations for present purpose regarding the structural analysis of disordered materials have been already provided in Chapter 4 and are not duplicated here.

(A) Oxide and Chalcogenide Glasses

The report on the structure of a germanate (GeO_2) glass by Bondot in 1974 may be referred to as the first example by applying the anomalous x-ray scattering to structural characterization of disordered materials. The near neighbor correlation functions of Ge-Ge, O-O and Ge-O pairs in a GeO_2 glass have been estimated from x-ray diffraction data in which the scattering ability of the constituents are varied by the anomalous dispersion effects on Cu-Kα and Ag-Kα radiations, coupled with the experiment by neutron diffraction. Here, the anomalous dispersion factors of Ge atoms used in his work are as follows; $f' = -1.31$, $f'' = 1.04$ for Cu-Kα and $f' = 0.38$, $f'' = 1.29$ for Ag-Kα. Note that these numerical values slightly differ from the theoretical dispersion factors reported by Cromer and Liberman (1970). The pioneer results of Bondot (1974) on a germanate glass are shown in **Fig.8.4**, although the experimental details are not available in his paper.

Gopalarkrishnan and Ramaseshan (1975) measured x-ray scattering intensities from a As$_{45}$Te$_{55}$ glass using Kα radiations of Mo, Cu and Cr, in order to separate the partial structure factors along the line proposed by Ramesh and Ramaseshan (1971). However, the results are still far from complete, mainly due to the rather limited wave vector range up to a value of 5.3 A^{-1}.

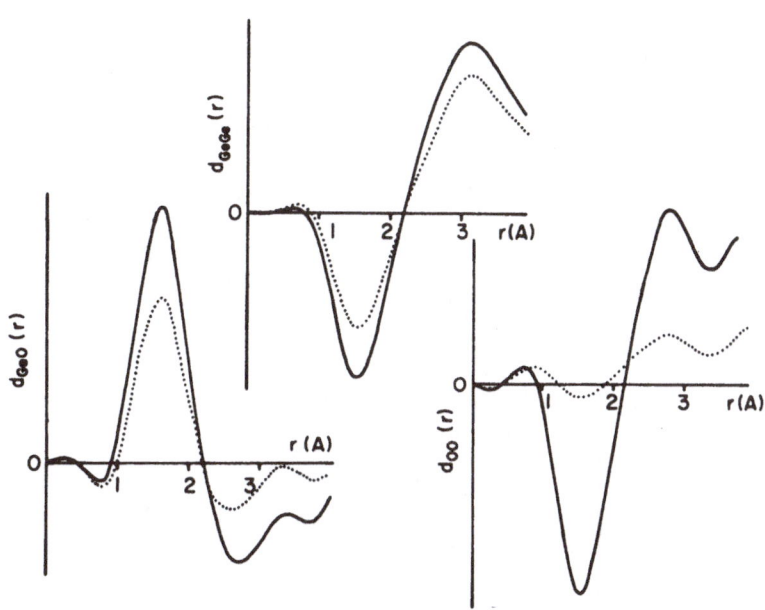

Fig.8.4 Partial radial distribution functions of GeO$_2$ glass in the near neighbor region. Direct experimental data (solid line) and values evaluated by the least-square analysis (broken line), (Bondot 1974).

(B) Metallic Alloy Glasses

As described earlier with information of Fig.5.1, the K-absorption edge of a Ni atom (8.332 keV = 1.488 A) is located near the energies (wavelength) of some characteristic Kα radiations produced from the commercial x-ray targets. For example, Cu-Kα (8.048 keV = 1.540 A), Ni-Kα (7.473 keV = 1.659 A) and Co-Kα (6.930 keV = 1.789 A). With this fact in mind, Waseda and Tamaki (1975, 1976) attempted to separate the partial structure factors in Ni-base disordered alloys

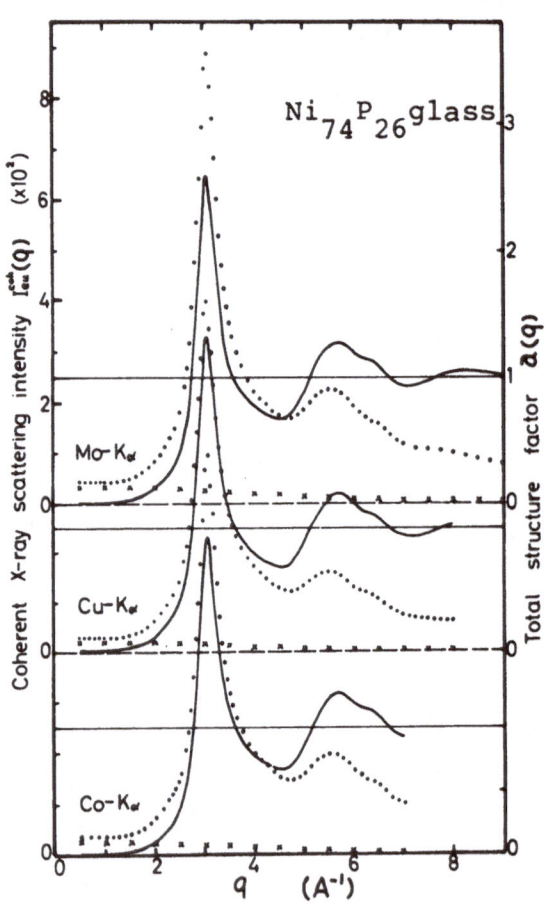

Fig.8.5 Coherent x-ray scattering intensity of the Ni$_{74}$P$_{26}$ glass obtained with three different Kα radiations (dotted line). Laue monotonic scattering intensity $<f^2>-<f>^2$ (cross) and total structure factor (solid line), (Waseda and Tamaki 1976).

from x-ray diffraction data alone, through the use of the anomalous (resonance) x-ray scattering. Their results are given below using the Ni-P glass as an example. Note that the anomalous dispersion factors in this alloy glass are expected to the first approximation as those of the pure element, because their electronic properties are known to show typical metallic character (Meisel and Cote 1977).

X-ray scattering intensity of Ni-P alloy glasses prepared by electrodeposition was measured at room temperature ($20^\circ C$) using three radiations of Mo, Cu and Co-Kα produced from the commercial x-ray targets. **Figure 8.5** gives the coherent x-ray scattering intensity per atom $I_a^{coh}(q)$ and the total structure factor $a(q)$ in the Faber-Ziman form of a $Ni_{74}P_{26}$ glass as an example. The crosses in this figure denote the so-called Laue monotonic scattering term ($<f^2> - <f>^2$). This has a significant effect in normalization of measured intensity data when the atomic scattering factors of the constituents differ from one another. As shown in Fig.8.5, the measured q-range is not so wide in the case of Cu and Co radiations. This gives rise to the source of error in normalization procedure. Note that the measurements with Mo-Kα radiation were done up to a sufficient high q-value of 17 A^{-1}.

In the literature, several normalization methods have been used for the structural analysis of disordered materials (see for example, Wagner 1972, Waseda 1980). Following the discussion and the results given in the previous works, the so-called Krogh-Moe-Norman's method may be recommended for normalizing measured intensity data for disordered systems (Krogh-Moe 1956, Norman 1957). This method is essentially based on the following equation.

$$\alpha_N \int_0^{q_{max}} q^2 [I^{obs}(q) - I^b(q)] dq = \int_0^{q_{max}} q^2 [<f^2> + I^{inc}(q) + I^{mul}(q)] dq$$

$$- 2\pi^2 \rho_o <f>^2 \qquad (8.6)$$

where α_N is the so-called normalization constant, ρ_o is the number density of atoms and $<f^2>$ and $<f>^2$ are the usual meanings. $I^{obs}(q)$ is the measured intensity data corrected by the polarization and absorption corrections, $I^b(q)$ the background intensity including the fluorescent radiation contribution and $I^{inc}(q)$ and $I^{mul}(q)$ are the incoherent and multiple scattering intensities which are possible to evaluate theoretically using the method proposed by Cromer and Mann (1967) and Malet et al. (1973). The fluorescent radiation contribution to the intensity becomes significant in the measurements with the energies near the absorption edge. Therefore, additional measurements for a crystalline sample having the same composition as that of the particular disordered materials is desirable for estimating the background intensity, because the fluorescent radiation from non-crystalline materials is equal to that of crystalline ones where the scattering intensity due to the atomic configuration is distinctly different in these two materials.

The total scattering intensity at higher q-value should be close to the term of $\phi(q) = [<f^2> + I^{inc}(q) + I^{mul}(q)]$ and thus the measure of the background intensity correction may be expressed by the sum of $I^b(q)$ and $\phi(q_{max})/\alpha_N$. Considering this point, we can obtain an additional guiding principle for normalization. However, the difficulty is still in determining the level of $I^b(q) + \phi(q_{max})/\alpha_N$ for the subtraction from $I^{obs}(q)$, because substantial oscillations in $I^{obs}(q)$ due to the atomic scale structure in disordered materials are generally found in the case where q_{max} is a relatively small value such as $q_{max} = 7.0$ A^{-1}. For this reason, an independent check for normalization proposed by Rahman (1965) is one way to reduce this difficulty. This Rahman's method has already been applied to several disordered materials, but the essential features of this useful method are given below, for future convenience.

The structure factor a(q) must satisfy the following relation which is readily derived from the physical significance of RDF = 0 for all values less than the diameter of the constituent atoms.

$$4\pi\rho_o L^3 \frac{J_1(\mu L)}{\mu L} = \frac{L}{\pi\mu} \int_0^{q_{max}} q[a(q)-1][J_o\{(q+\mu)L\} - J_o\{(q-\mu)L\}]dq \qquad (8.7)$$

where $J_i(x)$ is the i-th spherical Bessel function, μ is an arbitrary parameter with dimension of A^{-1} and L with dimension of A is considered to be less than the diameter of the constituent atoms. The calculation has to be done for a variety of values of μ and L in terms of the left-hand side of eq.(8.7). For example, these values for the $Ni_{74}P_{26}$ glass are listed in the third column of **Table 8.2** as ε_{exp}. The corresponding theoretical values evaluated from the left-hand side of eq.(8.7) with the measured density value (the experimental uncertainty in density less than 0.1 %) are also given in the forth column of this table as ε_{theor}. The magnitude of the error may be estimated as the correction factor by which the experimental data of a(q) must be multiplied to make the integrated value of ε_{exp} equal to that of ε_{theor}. Thus, the last column of Table 8.2 indicates the correction factor to obtain perfect agreement between these two values.

Table 8.2 Check of the normalization procedure for x-ray diffraction data of the $Ni_{74}P_{26}$ glass by the Rahman's method (Waseda and Tamaki, 1976).

radiation	L (A)	(A^{-1})	ε_{exp}	ε_{theor}	Correction factor
Mo-Kα	1.0	1.0	0.317	0.320	0.998
	1.0	2.0	0.251	0.253	0.989
	1.0	3.0	0.140	0.140	1.000
	2.0	1.0	1.883	1.880	0.992
	2.0	2.0	0.276	0.272	1.011
	2.0	3.0	-0.249	-0.267	1.003
Cu-Kα	1.0	1.0	0.322	0.320	0.992
	1.0	2.0	0.251	0.253	0.985
	1.0	3.0	0.136	0.140	1.005
	2.0	1.0	1.875	1.880	0.986
	2.0	2.0	0.270	0.272	0.991
	2.0	3.0	-0.263	-0.267	1.012
Co-Kα	1.0	1.0	0.315	0.320	0.988
	1.0	2.0	0.248	0.253	0.985
	1.0	3.0	0.143	0.140	1.010
	2.0	1.0	1.885	1.880	0.986
	2.0	2.0	0.267	0.272	1.012
	2.0	3.0	-0.261	-0.267	1.014

As shown in Table 8.2, the structure factor a(q) obtained for the $Ni_{74}P_{26}$ glass needs to be changed by less than 1 % to satisfy the relation of eq.(8.7) exactly for each μ and L. On the basis of these results, the normalization to obtain the so-called total structure factor is considered accurate with the uncertainty less than 1 % in the case of the $Ni_{74}P_{26}$ glass. We may also suggest the following point. When the difference between ε_{exp} and ε_{theor} appears to be significant in the measurements with the limited q-range, the normalization to obtain the structure factor is one way with which to provide the correction factors similar to those for the measurements with a sufficient wide q-range, although this modification is not with a sufficient wide q-range, although this modification is not applied to the results given in Fig.8.5 and Table 8.2.

Figure 8.6 shows the three partial structure factors evaluated from measured intensity data of the $Ni_{74}P_{26}$ glass given in Fig.8.5. Their general profile is similar to that of typical metallic glasses, i.e., the structure factor is characterized by small oscillations about unity after the usual first peak. However, it is found that the results are rather widely spread in certain positions. This is mainly due to the relatively small difference between the anomalous dispersion terms and the original atomic scattering factor, although

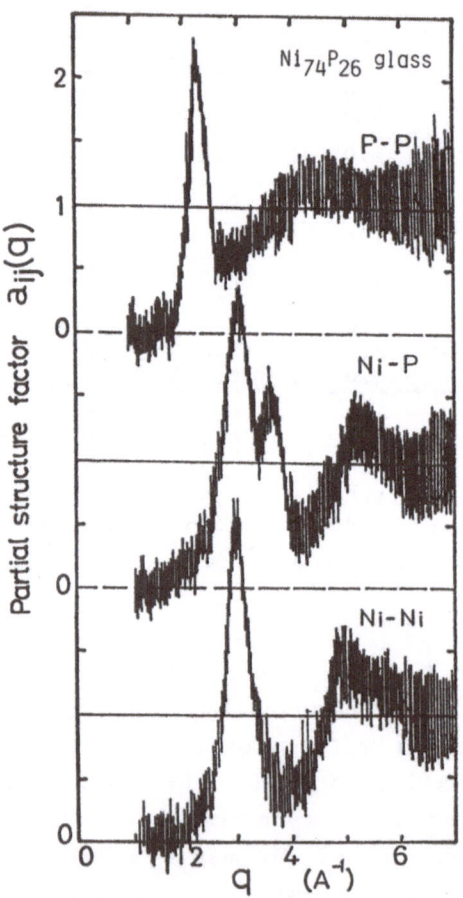

Fig.8.6 Partial structure factors of the $Ni_{74}P_{26}$ glass. The vertical lines denote the experimental uncertainty (Waseda and Tamaki 1976, Waseda 1981).

the variations detected are sizable as exemplified in **Table 8.3.** When the three partial structure factors, therefore, are evaluated by solving the three simultaneous equations based on eq.(4.8), the residual uncertainty is sometimes enhanced, and then only a physically meaningless solution can be obtained in some cases. However, the following suggestions may be worthy of note. A series tests using the computer technique have been carried out which confirm the significance of the numerical solution and of the principal features for the partial structure factors of Fig.8.6 (Waseda and Tamaki 1976) For example, the simultaneous equations were solved independently by the four methods (the Cramer, sweep-out, Gauss-Seidel and the combination of the sweep-out and Gauss-Seidel method in numerical calculation). Satisfactory agreement was found with respect to the numerical solutions. Moreover, an attempt was made to eliminate artificially the

Table 8.3 Anomalous dispersion terms and the experimental values of the total structure facors for $Ni_{74}P_{26}$ glass (Waseda and Tamaki, 1976).

Radiation		Mo-Kα	Cu-Kα	Co-Kα
Ni	f'	0.37	-3.20	-2.04
	f"	1.20	0.67	0.78
P	f'	0.09	0.28	0.33
	f"	0.10	0.43	0.57
q (A^{-1})		Total structure factor		
2.0		0.285	0.302(5.0)	0.293(2.8)
3.0		2.432	2.392(1.6)	2.414(0.8)
4.0		0.774	0.815(5.3)	0.791(2.2)
5.0		0.867	0.846(2.4)	0.852(1.7)
6.0		1.215	1.183(2.6)	1.201(1.2)
7.0		0.952	0.784(3.4)	0.978(2.7)

( ) Difference from the value of Mo radiation in the unit of %.

first peak from each partial structure factor in turn, but this modification provided gross disagreement between the observed and the
calculated intensity data. The elimination of the correlation of the
P-P pair also did not lead to a reasonable convergence between the
observed and the calculated intensities. Considering all these
factors, the results of Fig.8.6 are, in the author's view, rather
surprisingly good.

The partial structure factors of the $Co_{75}P_{25}$ glass have been also
evaluated using this anomalous x-ray scattering technique (Tamaki and
Waseda 1980). The results are shown in **Fig.8.7** together with those
obtained by the polarized neutron diffraction technique (Bletry and
Sadoc 1975) and by the combination technique of this polarized neutron
and usual x-ray diffraction (Sadoc and Dixmier 1976). In these three
sets of data, independently obtained by the different techniques, the
agreement is very encouraging, although there are differences in
detail.

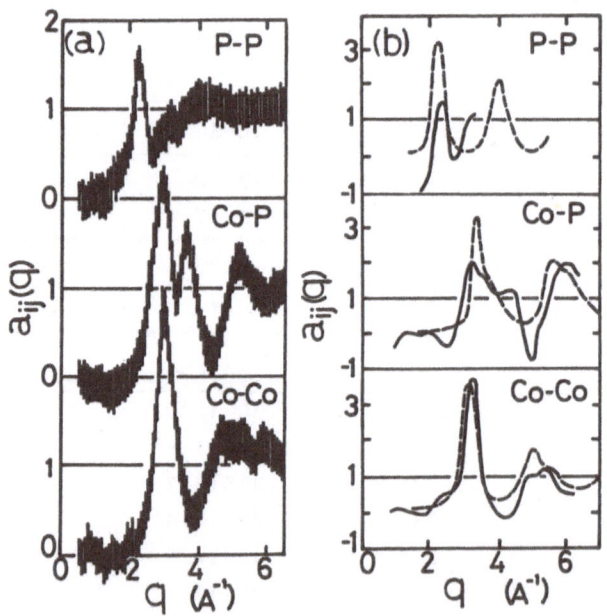

Fig.8.7 Partial structure factors of the $Co_{80}P_{20}$ glass.
(a) anomalous x-ray scattering (Tamaki and Waseda
1980). (b) solid line: polarized neutron scattering
(Bletry and Sadoc 1975), dashed line: x-ray and
polarized neutron scattering (Sadoc and Dixmier 1976).

The partial structure factors of the $Cu_{57}Zr_{43}$ glass determined by the anomalous x-ray scattering are illustrated in **Fig.8.8** (Waseda et al. 1977). Neutron diffraction with the isotope substitution technique (thus three different samples are required) has also been used to separate the partial structure factors of the $Cu_{57}Zr_{43}$ glass (Mizoguchi et al. 1978). As shown in Fig.8.8, the good agreement is again found in these two independent experimental results, however it may be noted that there are differences in detail.

As easily seen from the results of Figs.8.6-8.8, the structural study by the anomalous x-ray scattering is known to be restricted to the q-range up to about 7-8 A^{-1} in several cases and thus the spurious ripples in the resultant RDF due to the finite termination in the Fourier transformation through eq.(4.6) cannot be removed, even though the quite accurate and reasonable data of $a_{ij}(q)$ are employed. This is certainly one of the disadvantages in the structural investigation of disordered materials by the anomalous x-ray scattering with the common characteristic $K\alpha$ radiations. However, it is relatively easy to trace the positions where the spurious ripples due to the limited termina-

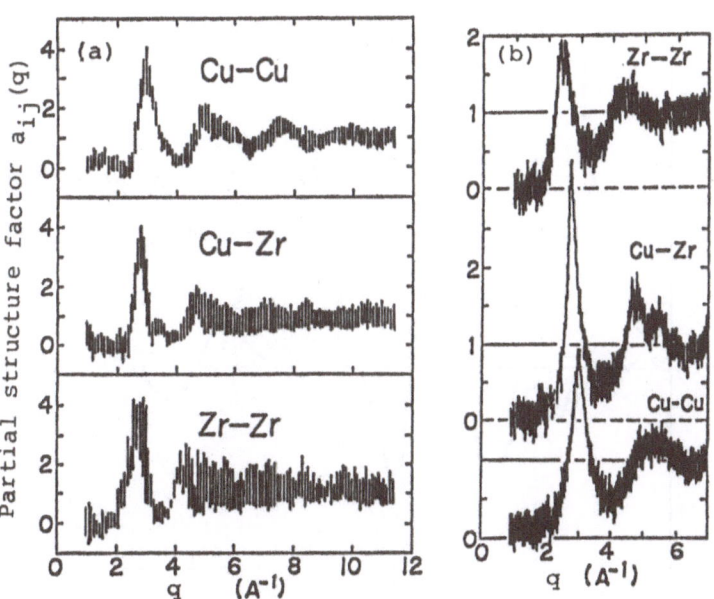

Fig.8.8 Partial structure factors of the $Cu_{57}Zr_{43}$ glass. (a) isotope substitution method of neutrons (Mizoguchi et al. 1978), (a) anomalous x-ray scattering (Waseda et al. 1977).

tion appear significant. For example, the oscillations of such ripples appear at $r = \pm 5\pi/2q_{max}$ or $\pm 9\pi/q_{max}$ from the principal peak position, where q_{max} is the upper limit of q obtained experimentally, following the works of Finbak (1949) and Sugawara (1951).

Figure 8.9 gives the partial RDF of P-P pairs in the $Ni_{74}P_{26}$ glass calculated from the corresponding partial structure factor of Fig.8.6 together with the result in the $Co_{81}P_{19}$ glass by Sadoc and Dixmier (1976) who used a combination technique of polarized neutron and x-ray diffraction. The arrows in this figure indicate the positions of the spurious ripples due to the termination effect in the Fourier transformation. On the basis of this trace, the peak observed at about $r = 2.0 A$ in both alloy glasses is considered the spurious one. It is not intended to describe here the detailed information on the structure of each alloy glass. However, the results of Fig.8.9 clearly suggest that phosphorus atoms unlikely occupy the nearest-neighbor positions in these alloy glasses, because the feasible first peak position (about 3.2-3.4 A) of the P-P pairs is considerably larger than the value (2.2 A) predicted from the atomic size of a P

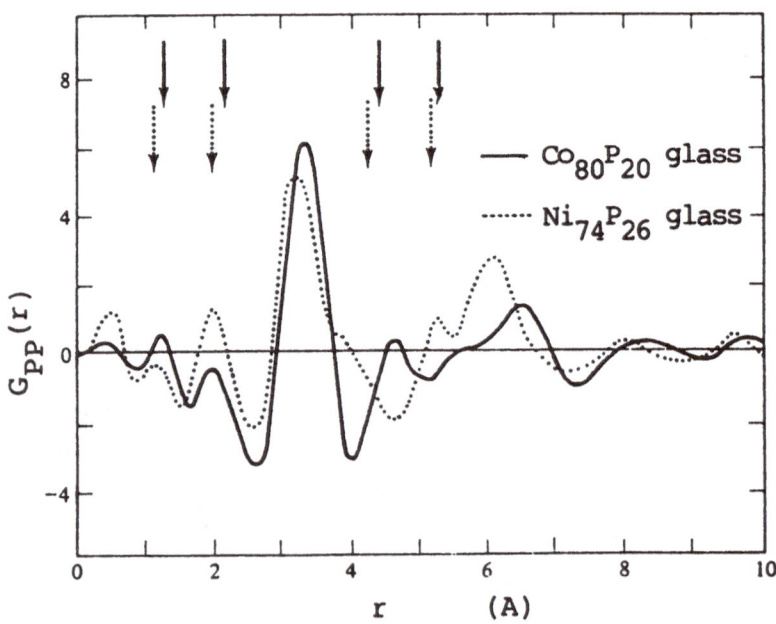

Fig.8.9 Partial radial distribution function G(r) of P-P pairs in metallic alloy glasses of $Co_{80}P_{20}$ (Sadoc and Dixmier 1976) $Ni_{74}P_{26}$ (Waseda and Tamaki 1976). The arrows indicate the positions of spurious ripples due to the termination effect.

atom. This is well-known to provide a great impact on advancing the quantitative interpretation of the structure for various metal-metalloid type alloy glasses.

In the analysis by applying the anomalous x-ray scattering, the resultant partial structure factors are frequently dispersed in places as shown in Fig.8.6. This is mainly attributed to the relatively small variation (less than 10 %) due to the anomalous dispersion effect on the normal atomic scattering factor, so long as one can use the common $K\alpha$ radiations produced by commercial x-ray targets. In order to reduce such difficulty, the refining sequence as exemplified by **Fig.8.10** is frequently used so as to satisfy some physically meaningful conditions.

For example, the measured intensity data must always be positive and the general profile of the structure factors, which are smooth functions without sharp changes of slope near the first peak region, seems to be feasible on the basis of the structural analysis for various disordered materials (see for example, Furukawa 1962, Waseda 1980). A similar refining method has also been employed in the structural study of molten salts by the isotope substitution technique for neutrons (Edwards et al. 1975, Mitchell et al. 1976). The following supplemental data processing is also helpful, when the measured intensity data with different compositions over a certain composition range are available.

As described in Chapter 3, the partial structure factors, in principle, depend on the composition. However, the change in the total structure factors of disordered materials is frequently **gradual and monotonic in a certain composition range.** In this regard, the concentration dependence of partial functions seems to be **not so severe in disordered materials over a limited concentration range** and the assumption of the concentration independence for the partial structure factors may be acceptable as a first approximation. This assumption was originally introduced by Halder and Wagner (1967) to evaluate the partial structures in liquid Ag-Sn alloys and later applied to many metallic alloy liquids.

With this fact in mind, the three partial structure factors, for example, of the Cu-Zr alloy glasses can be evaluated, initially, from measured intensity data with 35, 40 and 45 at % Zr obtained by a single radiation such as Rh-Kα by using the so-called Halder-Wagner's scheme assuming the concentration independence of partial functions. The Cu-Zr alloy glasses with a small composition range (35-45 at % Zr)

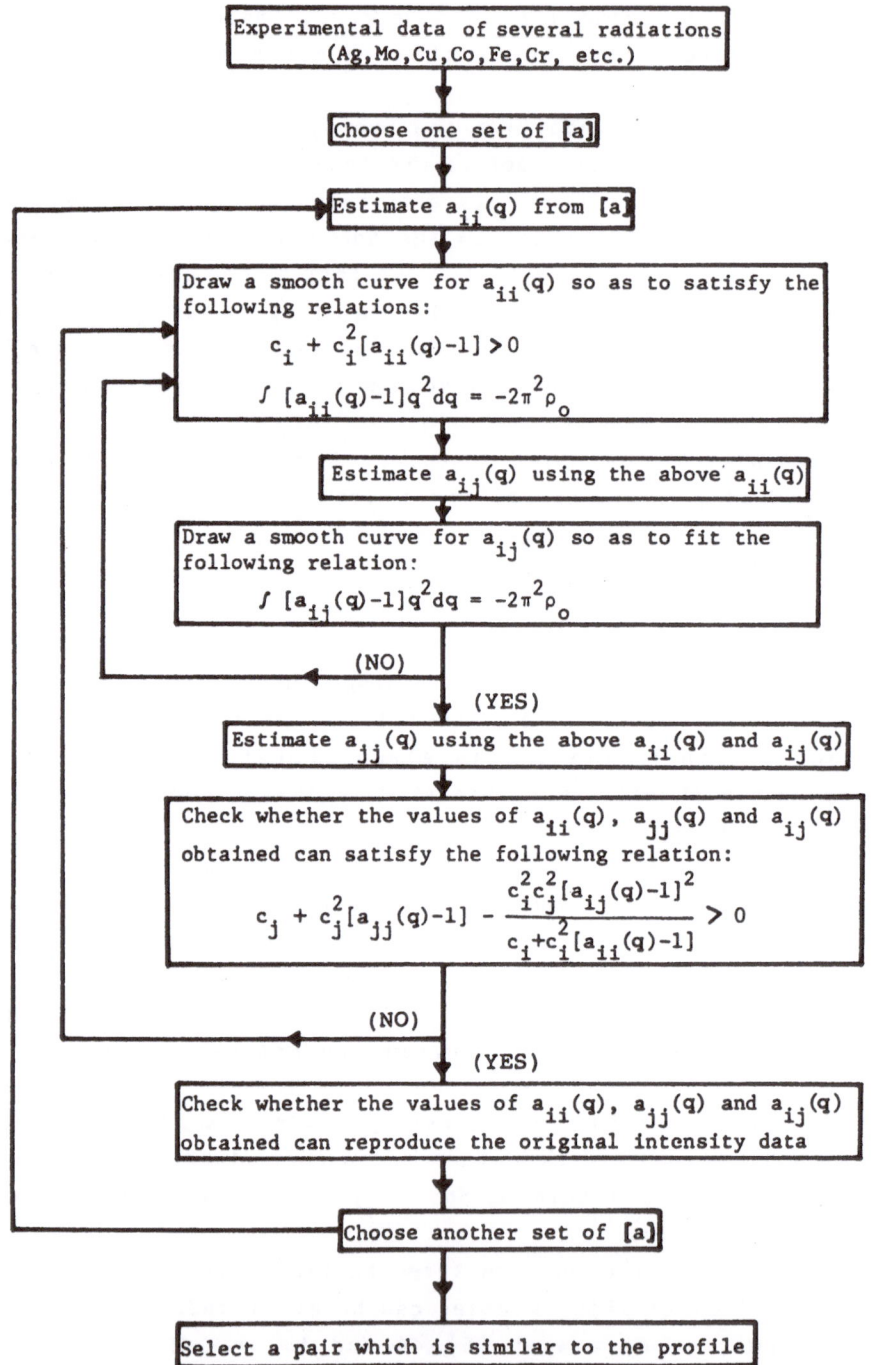

Fig.8.10 Refining sequence for analysis of measured intensity data
         by the anomalous x-ray scattering (Chen and Waseda 1979).

might be in a sense of the Halder and Wagner's approach (1967). The coherent scattering intensity from these partial structural data obtained in the first step and the Laue monotonic scattering term are calculated with the anomalous scattering contributions and then compared with the experimental data obtained by other $K\alpha$ radiations such as Cu and Fe. In this second step, the refining sequence of Fig.8.10 is also employed. When such supplemental data processing indicates that the systematic difference of the order of 5-10 % detected among the experiments with different characteristic radiations can be explained by the different components of anomalous x-ray scattering from constituent atoms, the partial structure factors evaluated in the first step using the Halder-Wagner's scheme are considered to be known at least in a sense of **the necessary condition at best**, although they might be **not the sufficient condition**. It may be noted that the conceptional appraoch employed in this data processing, in which the difference observed in intensity with different energies (radiations) is attributed to the anomalous dispersion contribution, was originally suggested by Hosoya (1970) and later by Shevchik (1977). This approach has been successfully applied recently to the structural study of the Ge-Se glass by Fuoss et al. (1981) and of the Mo-Ni glass by Aur et al. (1983a) coupled with the synchrotron radiation source. In addition, this type of supplemental data processing for the anomalous x-ray scattering with the help of the Halder-Wagner's scheme is one way to reveal much valuable information on the structure of disordered materials, in cases where other methods such as the isotope substitution technique of neutrons are found to be technically difficult or until the full potential of the anomalous x-ray scattering itself can be assessed as a reliable and established tool for evaluating the partial functions of various disordered materials.

(C) Metallic Alloy Liquids

The L absorption edge of Ce atoms (L_{III} = 5.725 keV, L_{II} = 6.161 keV, L_I = 6.560 keV) is also located near the energies of conventional x-ray sources such as Cr-$K\alpha$ (5.415 keV), Fe-$K\alpha$ (6.404 keV) and Co-$K\alpha$ (6.930 keV). On the other hand, Ce is known to exhibit interesting magnetic properties on alloying with transition metals such as Ni and Co. For example, in alloys with magnetic elements such as Co, the magnetic susceptibility χ_m of liquid Ce-Co alloys varies nonlinearly with concentration and indicate a minimum at about equi-atomic composition (Schlapbach 1974). This behavior has been interpreted by postulating a change in electron occupation number of the 4f-level in these alloy liquids (see for example, Avigon and Falicov 1974). In this regard, one can expect some different structural behavior on alloying. This contrasts with the case of Ce-Cu alloys in the liquid state, since only simple linear concentration dependence is found in the magnetic susceptibility (Schlapbach 1974).

Figure 8.11 shows the partial structure factors for liquid Ce-Co alloys from measured intensity data of two different compsotions (25 and 75 at% Co) by using the anomalous x-ray scattering (Waseda and Toguri 1978). Similar results of liquid Ce-Cu alloys are given in Fig.8.11 for comparison. The vertical lines in this figure denote the residual experimental uncertainty and these are of the order of ± 0.3. The following important observation could be obtained from these results.

There is a significant difference between the structure factor of pure liquid Ce and the partial structure factor of the Ce-Ce pairs for a Co-rich (75 at% Co) alloy. This implies that the effective interaction distance of Ce-Ce pairs in liquid Ce-Co alloys depends upon the Co content, since the Ce atoms in Co-rich region interact more closely than in Co-dilute region. In contrast, the liquid Ce-Cu alloys do not show such behavior i.e., the partial structure factor of Ce-Ce pairs for the Ce-Cu case is found to be insensitive to a change in composition.

The detailed discussion on this subject is out of the scope of this article: however the structural information obtained by applying the anomalous x-ray scattering suggests that some of the f-electrons in the Ce atoms might be transferred to the 3d-states of the Co atoms on alloying. This inference is also confirmed by the lack of similar structural behavior in liquid Ce-Cu alloys, because the contribution from the d-shell in Cu atoms is much smaller than that in Co atoms. A relationships between electron localization and atomic scale structure in semiconducting liquids such as III_b(Ga, In, Tl)-Te alloys has also been confirmed by the partial structures evaluated by the anomalous x-ray scattering technique (see for example, Waseda 1981, Takeda et al. 1983).

69

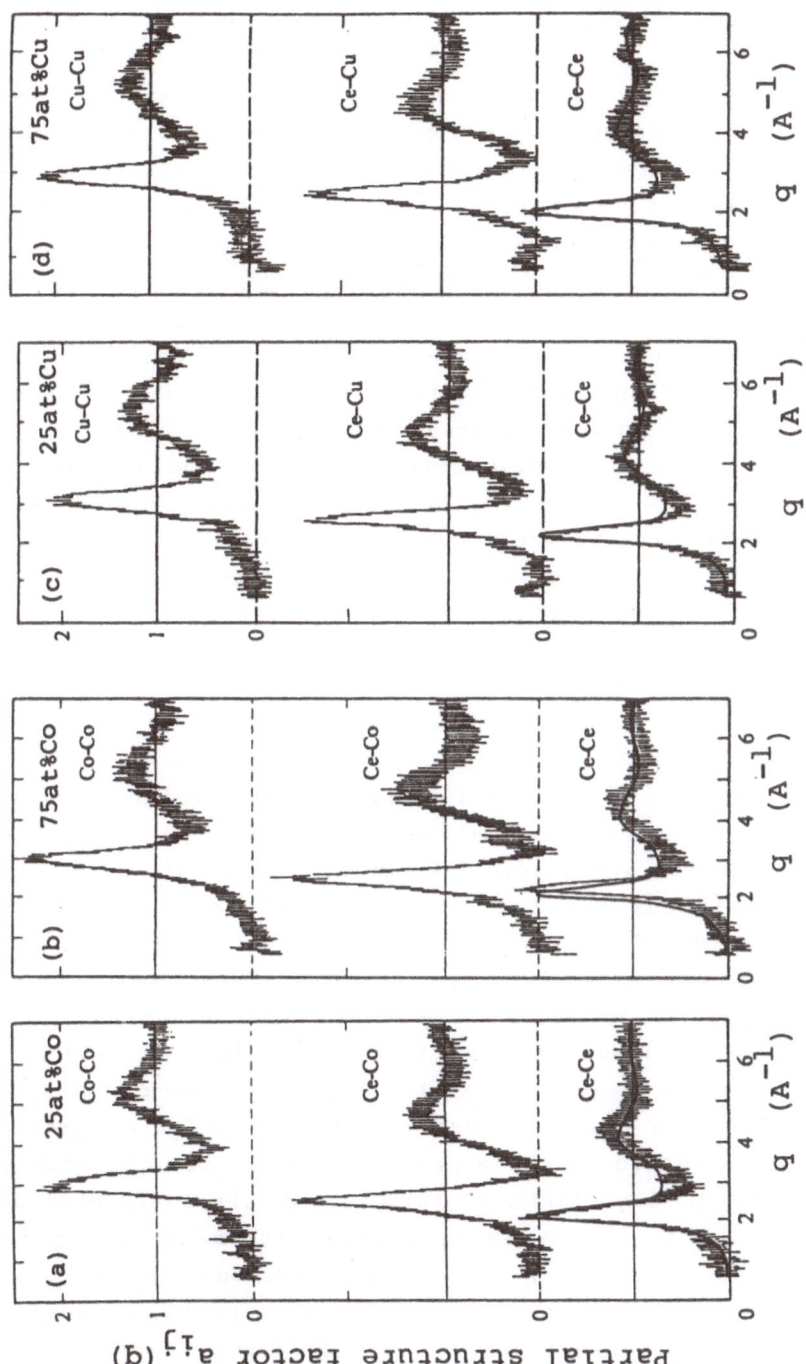

Fig.8.11 Partial structure factors of liquid Ce-Co and Ce-Cu alloys at about 20°C above the liquidus temperature. (a) Ce75Co25, (b) Ce25Co75, (c) Ce75Cu25, (d) Ce25Cu75 . The solid lines denote the structure factor of liquid pure Ce at 870°C (Waseda and Toguri 1978).

(D) Solid Fast Ion Conductors

The relatively new results on solid fast ion conductors may also be described as one of the good examples obtained by the anomalous dispersion effect of x-rays with respect to the structural characterization of disordered materials. Solid fast ion conductors such as noble metal chalcogenides and noble metal halides have been recently drawing attention, because of their unusually high ionic conductivity in the high temperature phase at about $200^{\circ}C$ (see for example, Mahan and Roth 1976, Hagenmuller and van Gool 1978). The ionic transport properties of the noble metal solid fast ion conductors such as Ag_2S and AgI imply that the cations in these solid fast ion conductors are almost in the liquid-like disordered state, although they are macroscopically still in the typical solid state, because their tracer-diffusion coefficient was found to be comparable to the self-diffuion coefficient of liquid metals (see for example, Okazaki 1967). However, the previous structural studies had relatively little impact on interpreting their unusual properties. The use of the anomalous x-ray scattering has recently brought a significant break through in this subject (see for example, Tsuchiya et al. 1978, 1979). The essential points are given below.

The solid fast ion conductors indicate the particular x-ray diffraction patterns characterized by the coexistence of both crystal-like (Laue-Bragg) peaks and liquid-like diffuse patterns, as exemplified by **Fig.8.12** using the results of a high temperature phase (hereafter referred to as α-phase) of AgI as an example. The Laue-Bragg peaks, of course, are mainly attributed to the periodic array of anions first suggested by Stock in 1934. A few diffraction studies concerning the liquid-like diffuse patterns have been reported, but the results are not sufficient enough to allow a reliable interpretation, probably because the contribution from anion-cation pair has been neglected ab initio in these previous works (see for example, Sakuma et al. 1977, Hoshino 1978). In this regard, the newly developed theoretical equations for x-ray scattering intensity of systems showing both Laue-Bragg peaks and diffuse patterns lead to the contributions from both anion-cation and cation-cation pairs, without any assumption for the distribution of cations (Tsuchiya et al. 1978, 1979). It is not the author's intention here to discuss the theoretical background, in detail, on the x-ray scattering intensity for solid fast ion conductors. Such information is available in **Appendix 3**. We give here only one example on the application of the newly developed theoretical equation to analysis of measured intensity data of α-AgI, in order to demonstrate the usefulness of the anomalous x-ray scattering technique.

An x-ray scattering intensity of α-AgI given in Fig.8.12 indicates a clear distinction between the crystal-like Laue-Bragg peaks and the liquid-like diffuse pattern. Although these peaks can be easily coincident with the bcc lattice formed by iodine atoms by the conventional crystallographic method, their relative intensities are inconsistent with those expected for a typical bcc structure as shown in **Table 8.4**. This suggests that the coherent scattering from cation-anion and also cation-cation pairs contributes partially to these Laue-Bragg peaks, as discussed in detail by Tsuchiya et al. (1979). Our intention here is concerned rather with the liquid-like diffuse patterns.

The liquid-like diffuse components $S^{dif}(q)$ obtained from measured intensity data by subtracting the crystal-like Laue-Bragg peaks are expressed by the following equation (Tsuchiya et al. 1979, see also Appendix 3).

$$S^{dif}(q) \propto \frac{f_A f_B^* + f_A^* f_B}{x f_A f_A^* + (1-x) f_B f_B^*} x(1-x) [\tilde{S}_{AB}(q)-1] + \frac{(1-x)^2 f_B f_B^*}{x f_A f_A^* + (1-x) f_B f_B^*} [\tilde{S}_{BB}(q)-1] \qquad (8.8)$$

where x is the concentration of anions and A and B denote anion and cation, respectively. That is, the diffused intensity contains interference terms of both anion-cation and cation-cation pairs, $S_{AB}(q)$ and $S_{BB}(q)$. It is also noteworthy that the expression for the weighting factors of two partial functions is the same as the one given for liquid binary alloys (Faber and Ziman 1965, see also Chapter 3) and thus the observed diffuse patterns can be decomposed into anion-cation and cation-cation contributions by the standard technique including the anomalous x-ray scattering.

The anomalous dispersion effects from the constituent elements yield a significant difference (the order of 5 %) in the experiments of α-AgI. The numerical examples of the values for $S^{dif}(q)$ at 2.50 A^{-1} are 1.183 for Mo-Kα and 1.262 for Cu-Kα radiations, respectively (Tsuchiya et al 1979). The two partial structure factors of Ag-I and Ag-Ag pairs estimated from measured diffuse intensity patterns are given in **Fig.8.13**. The vertical lines denote the fluctuations due to

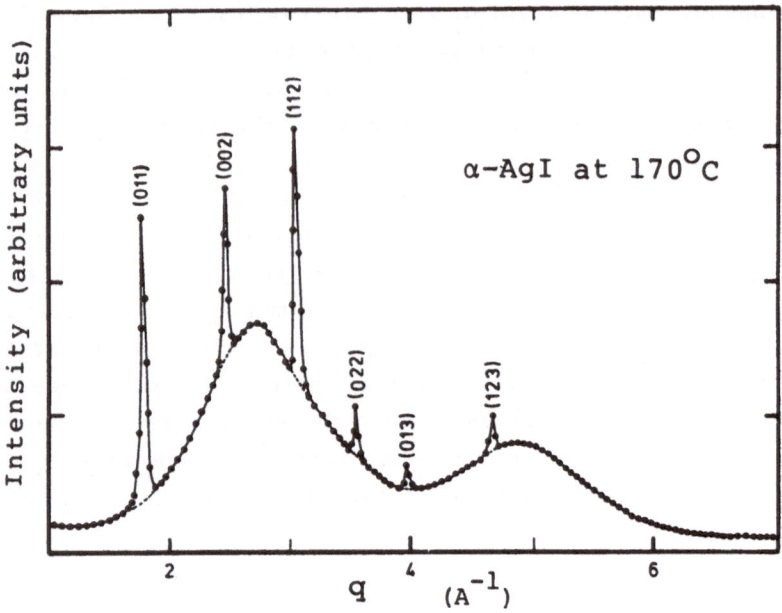

Fig.8.12 X-ray scattering intensity of αAgI at 170°C
(Tsuchiya et al.1979).

Table 8.4 Experimental data of Debye lines for α-AgI at 170°C
(Tsuchiya et al., 1979).

d(A)	(hkl)	I/I_1	typical bcc structure (W-powder)
3.586	(110)	100(I_1)	100
2.533	(200)	63	15
2.069	(211)	85	23
1.792	(220)	12	8
1.602	(310)	4	11
1.463	(222)	—	4
1.357	(321)	9	18
1.267	(400)	—	2

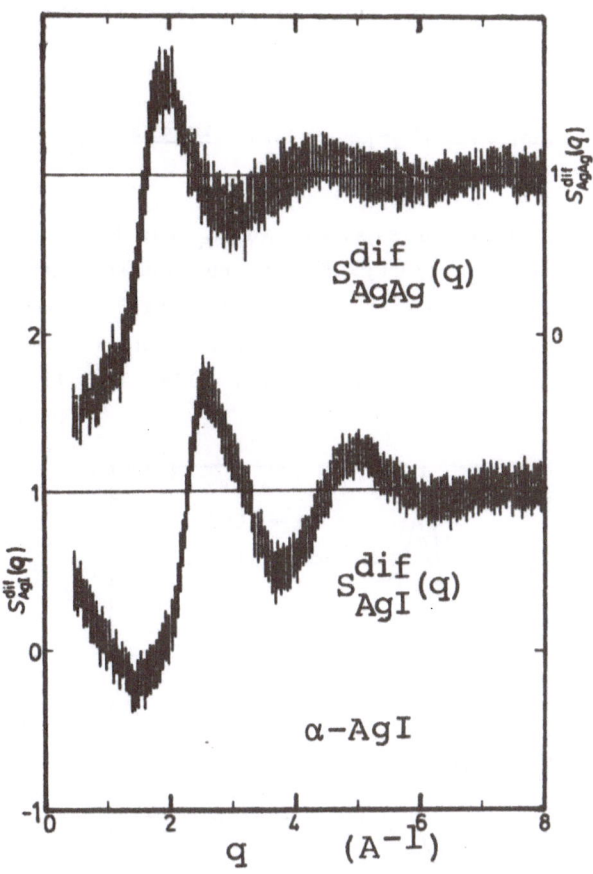

Fig.8.13 Partial structure factors in liquid like diffuse
pattern of α-AgI at 170°C. (Tsuchiya et al.1979).

the experimental uncertainty as discussed previously in the case of disordered metallic alloy systems. Although these are of the order of ± 0.2, the basic profile of $S_{AgI}(q)$ and $S_{AgAg}(q)$ clearly appears to be characterized in a similar way in metallic liquids. In particular, $S_{AgAg}(q)$, similar to those of typical metallic liquids, directly implies that the distribution of the Ag ions in α-AgI are in a highly disordered state almost resembling a liquid-like one. This is reasonably consistent with the tracer-diffusion experiments reported by Okazaki (1967).

The average information on the atomic scale structure as a function of the relative distance between anion and cation or cation and cation can be obtained by the Fourier transformation in the following equation.

$$\tilde{g}_{ij}(r) = 1 + \frac{V}{2\pi^2(N_A + N_B)} \int_0^\infty q^2[\tilde{S}_{ij}(q) - 1]\frac{\sin(q \cdot r)}{q \cdot r} \, dq \qquad (8.9)$$

where V is the volume of a system and N_A and N_B are the number of anions and cations, respectively. The corresponding information on the atomic scale structure of α-AgI is shown in **Fig.8.14** together with the distribution function of anion-anion pairs. The results obtained by molecular dynamics (Vashishta and Rahman 1978) is also illustrated in

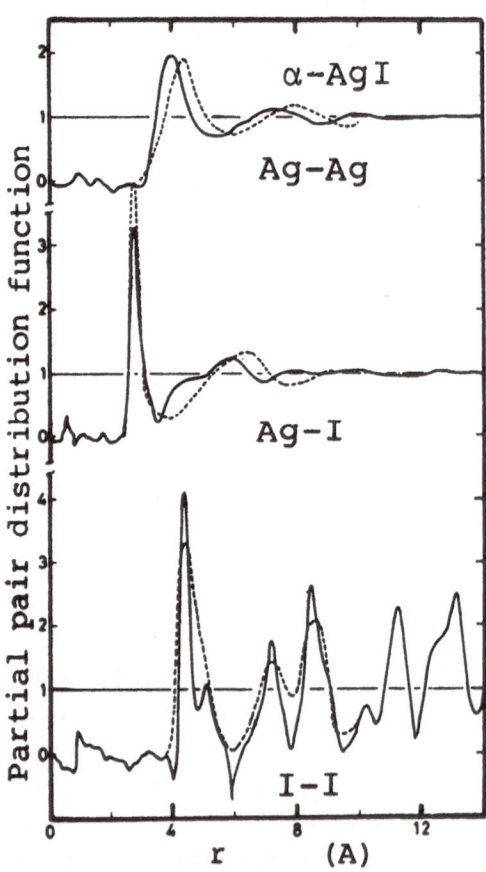

Fig.8.14 Partial distribution functions for three components in α-AgI at 170°C (Tsuchiya et al. 1979). The dotted lines are the results by molecular dynamics (Vashishta and Rahman 1978).

Fig.8.14 for comparison. As shown in this figure, the results obtained by the anomalous x-ray scattering essentially agree with those of molecular dynamics, although the phase difference between experimental and calculated $g_{AgAg}(r)$ is somewhat large. As shown in Fig.8.14, $g_{AgI}(r)$ oscillates nearly out of phase with $g_{AgAg}(r)$ and thus the charge cancellation in α-AgI is quite feasible as in the case of molten salts (see for example, Mitchell et al. 1976).

The average interionic distance between silver ion and iodine ion estimated from the partial radial distribution function data in Fig.8.14 is 2.75 A. This value is about 3 % smaller than the value of 2.83 A expected in the case when a silver ion occupies the center of the tetrahedral site with the lattice constant of 5.07 A. The recent EXAFS data reported by Hayes et al (1978), indicates that the near-neighbor Ag-I correlation is divided into two Gaussian peaks with r = 2.77 A and r = 2.93 A. The resolution of the present anomalous x-ray scattering results is still limited, in comparison with the EXAFS technique, so that these two possible near neighbor correlation distances may not be distinguished. However, the present author maintains the view that the results by the anomalous x-ray scattering are not inconsistent with those of EXAFS data and reveal much valuable information, because the EXAFS results (Hayes et al. 1978) provided only the near neighbor correlation of Ag-I pairs, whereas the anomalous x-ray scattering results give the RDFs of both Ag-I and Ag-Ag pairs as a function of distance. Other details on the particular structure of solid fast ion conductors may referred to in the original papers or recent review articles (see for example, Tsuchiya et al. 1978, 1979, Hoshino 1978, Boyce and Huberman 1979, Waseda 1981).

8.3 Recent Experimental Results by the Anomalous X-ray Scattering with a Synchrotron Radiation Source

The selected examples presented here indicate the effectiveness of the anomalous x-ray scattering for determining the fine structure of disordered materials. However, as long as we use only character-istic $K\alpha$ radiations produced by commercial x-ray targets, the change arising from the anomalous dispersion effect is not larger than about 10 %, because the energies of these $K\alpha$ radiations are often not close enough to the absorption edges of the sample constituents. For this incidental problem, the previous results are qualitatively well recognized, but their quantitative accuracy is often accepted with some reservations.

With respect to this point, the use of the energy dispersive x-ray diffraction (EDXD) technique (see for example, Egami 1981b and **Appendix 2**) is found to give in reducing this difficulty by making use of the energy range in which the anomalous dispersion effect is well detected. For example, when studying Pd alloy glasses with the conventional angular scanning technique, only the characteristic radiation of Ag-Kα (22.165 keV) is available near the K-absorption edge of Pd atoms (22.348 keV) in which the anomalous dispersion effect is not so obvious, because the energy difference exceeds 2000 eV. On the contrary, the EDXD technique with the solid state detector covers a wide

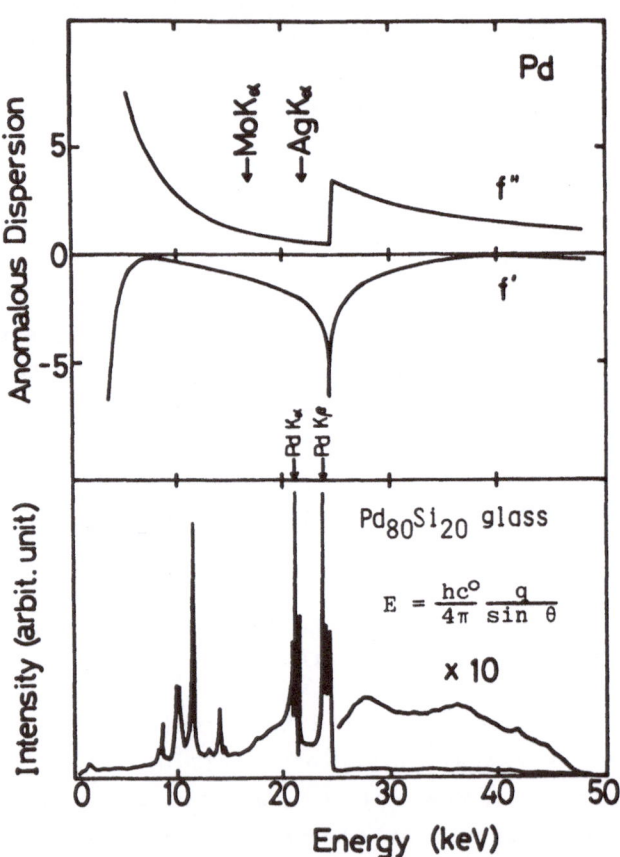

Fig.8.15 Anomalous dispersion factors f' and f" of Pd atoms and intensity pattern of the Pd$_{80}$Si$_{20}$ glass obtained by EDXD. Peaks below 10 keV are L lines of W and fluorescent radiations of impurity (Egami et al. 1978).

range of energy in which the anomalous x-ray scattering contribution is appreciable as exemplified in **Fig.8.15** using the preliminary results of the $Pd_{80}Si_{20}$ glass.

Since the relation between the energy E and the wave vector q is given by $q = (4\pi/hc^o)\sin\theta\cdot E$ (see eq.(2.3)), the angular scanning measurement coupled with the energy dispersive mode by using the solid state detector (SSD) provides the necessary items required to evaluate the partial functions by the anomalous dispersion effect of x-rays. Such experimental results given in **Fig.8.16** clearly indicate the sizable difference (of the order of 10 %) in the scattering intensity at the two different energies of 17.4 and 24 keV, although the data

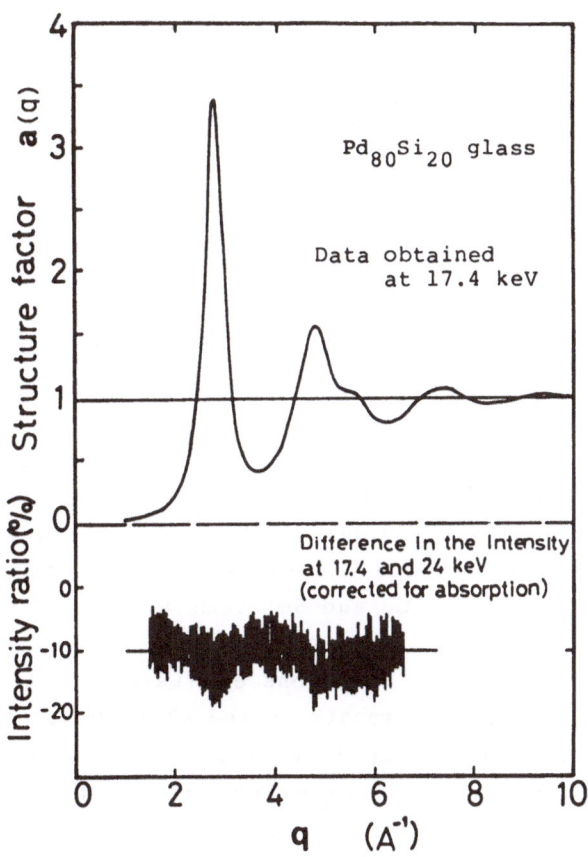

Fig.8.16 Difference in x-ray scattering intensity data at two different energies (17.4 and 24 keV) obtained by the angular scanning method with the energy dispersive mode (Waseda 1981).

processing for these experiments is not yet completed (Waseda 1981).

These encouraging results again strongly suggested that the use of a intense white x-ray source such as the synchrotron radiation could provide much successful information on the partial structures for various multi-component disordered materials through the **anomalous dispersion effect.** One of the main advantages in the experiment with the synchrotron radiation source is to tune the energy of the incident x-rays to an appropriate energy in the close vicinity of the absorption edge of the sample constituents where the anomalous disper- sion factors are very large and thus a significant improvement in the accurate determination of the partial structural functions can be achieved. In spite of this merit, very few applications are reported, because even though the principle is rather simple and straightforward than other techniques such as the EXAFS, the actual implementation of the anomalous x-ray scattering is not a trivial task. We give here the presently available information on the structural characterization of disordered materials by the anomalous x-ray scattering technique with the synchrotron radiation source.

Fuoss et al. (1980) have reported the results of the three partial structural functions of a GeSe glass evaluated from the measurement with the synchrotron radiation source at Stanford (California, U.S.A.). The anomalous dispersion factors were also determined by the combination technique of the absorption measurement (for f'') and the dispersion relation (for f') along the line proposed by Kawamura and Fukamachi (1978) and the results are illustrated in **Fig.8.17** using the Ge K-edge case. The measurement of x-ray scattering intensity was carried out at energies of 11.095, 11.105, 12.650 and 10.600 keV which were selected by a two-crystal (parallel setting) monochromator. The K-absorption edges of the Ge and Se atoms are 11.103 and 12.655 keV, respectively.

The partial structural functions evaluated are given in **Fig.8.18** and these results clearly demonstrate the usefulness of the **anomalous x-ray scattering** for structural study of disordered materials, when coupled with the **synchrotron radiation source.** However, it should be remembered that some reservations have also been stressed regarding the quantitative accuracy of their own results in Fig.8.18, because of the scarcity of beam time and some experimental difficulties in that time (Fuoss et al. 1980).

As easily seen from the results in Fig.8.17, the particular near- edge phenomena such as the white line is obvious in the Ge K-edge.

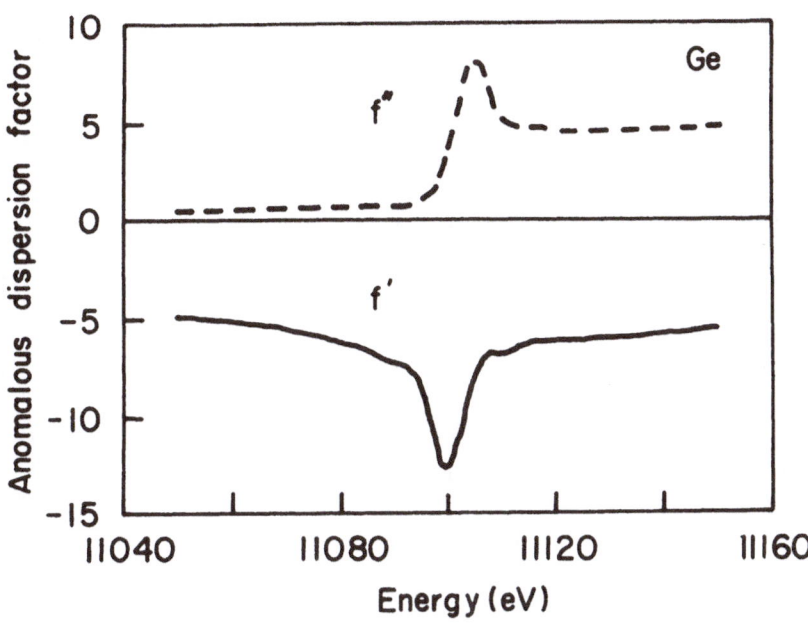

Fig.8.17 Anomalous dispersion factors of Ge atoms in the GeSe
        glass determined by the absorption measurement (f")
        and the dispersion relation (f'). E_K=11.103 keV for
        Ge atoms (Fuoss et al. 1980).

Since the energies employed are 2 - 8 eV away from the Ge edge (5 eV
away from the Se K-edge), the fluorescent radiation should be significant. Therefore, the diffracted beam monochromator of a flat LiF crystal and a scintillation counter employed in their measurement may not
give the sufficient elimination of the fluorescent component from
measured intensity data, in that time, in order to allow the evaluation of reliable information. The higher harmonic component of the
incident beam is also known to sometimes prevent exact tuning of the
monochromator crystal. The energy and intensity from a monochromator
crystal is not always stable in the synchrotron radiation source,
because the first monochromator crystal is heated by the higher
intensity x-ray flux and changes its lattice constant and then moves
in the anharmonic mode with the rocking curve of the second monochromator crystal. For these reasons, the results of Fig.8.18 reported by
Fuoss et al. (1980) are very impressive, but may be too sensitive to
the change in the anomalous dispersion effect which responds to the
incident x-ray energy. In this regard, **the energies used in the
structural study by the anomalous x-ray scattering technique may be
not too close to the absorption edge, less than about 30 eV,** in order

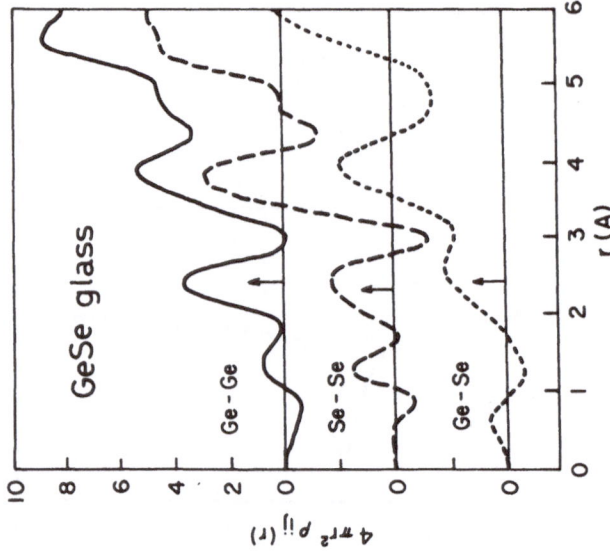

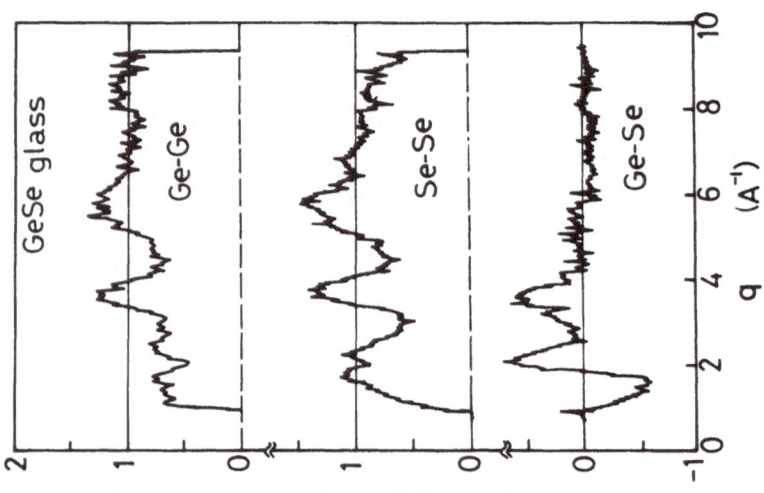

Fig.8.18 Partial structural functions of the GeSe glass evaluated by the anomalous x-ray scattering with a synchrotron radiation source at energies of 11.095, 11.105, 12.650 and 10.600 keV. $E_K=11.103$ keV for Ge atoms and $E_K=12.655$ keV for Se atoms (Fuoss et al.1980).

to avoid the uncertainty in the anomalous dispersion factors mainly arising from the near-edge phenomena, as already suggested by Hosoya (1970).

On the basis of the selected examples mentioned here, the usefulness of the anomalous x-ray scattering technique is well recognized, particularly with the synchrotron radiation source. However, it is also noticed that further developments are required before the full potential of this technique can be assessed. One of the major experimental difficulties in the anomalous x-ray scattering is the elimination of the fluorescent radiation from measured intensity data. Since the synchrotron radiation source is unstable over a long period of time and in the very noisy environment, a careful monitoring of the incident beam intensity should be also required.

Energies below the absorption edge are usually employed for the anomalous x-ray scattering experiment to prevent strong fluorescent radiation from the sample. However, even in such cases, there is still some fluorescent component excited by the two different mechanisms; one is **by the higher harmonic diffraction** from the monochromator crystal and the other is **due to the tail of the band pass** of the monochromator when it is tuned close to the absorption edge. The fluorescent radiation intensities from both of these origins change with time as the monochromator crystal is heated up or as the electron path shifts in the storage ring.

With respect to these experimental requirements, the use of an energy sensitive solid state detector (SSD) appears to be one way to provide the answer for questions un-solved by the usual scintillation or proportional counter. Of course, the fluorescent radiation can be discriminated by the diffracted beam crystal analyzer (Fuoss et al 1980, 1981), but this results in a severe reduction in intensity and thus decreases the statistical accuracy.

On the other hand, the typical energy resolution of about 300 ev at 20 keV of a SSD can discriminate most of the fluorescent radiation and also minimize the sensitivity to the high harmonic component of the incident x-rays. **Figure 8.19** shows a typical energy spectrum obtained by a intrinsic Ge detector for the measurement of a $Mo_{50}Ni_{50}$ glass (Aur et al. 1983b). Even though the energy of incident x-rays was tuned at 19.478 keV, far below the K-absorption edge (20.004 keV) of Mo atom, the fluorescent lines were quite appreciable. As easily seen from the spectrum in Fig.8.19, energy resolution of a SSD is high enough to separate clearly the Mo-Kα radiation from the elastically or

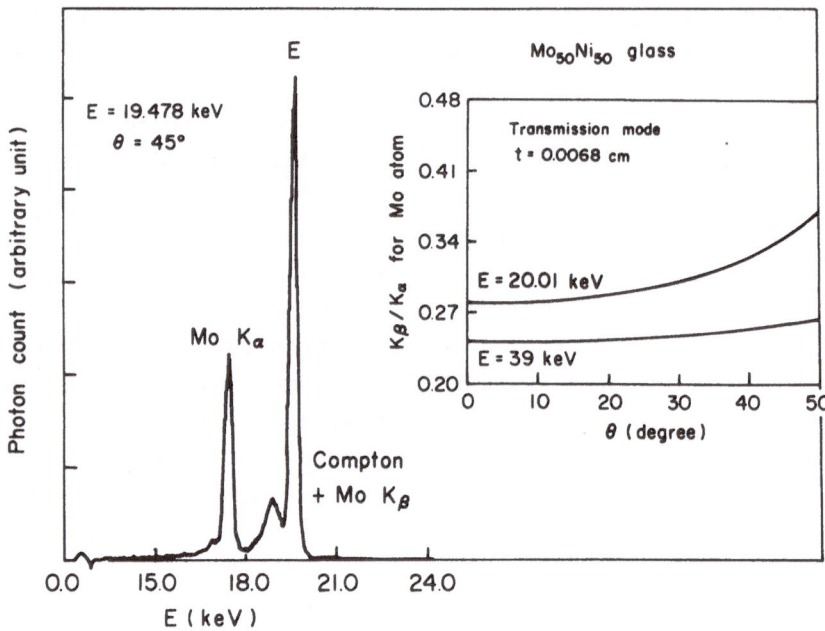

Fig.8.19 Typical spectrum of scattered x-ray photons in the
measurement for the $Mo_{50}Ni_{50}$ glass near the K edge of
Mo atoms by the Ge-SSD. The Kβ/Kα ratio calculated for
the tail of monochromator band pass (E=20.01 keV) and
for the second harmonics (E=39 keV), (Aur et al.1983b).

inelastically (Compton) scattered photons in a very electrically noisy
environment of the synchrotron radiation laboratory. Whereas,the Kβ
component could not be separated by a SSD, although the Compton component can be evaluated by the theoretical values (see for example,
Cromer and Mann 1967) with the so-called Breit-Dirac recoil factor.
However, the Kβ radiation can be numerically subtracted in the data
reduction process, so long as the Kα radiation is recorded by a single
channel analyzer (SCA) during the course of the experiment. The
intensity ratio of Kβ/Kα is relatively easy to determine through the
so-called off-Bragg condition or through the theoretical calculation
(Aur et al. 1983b). It should be noted, however, that this intensity
ratio depends upon the scattering angle and the diffraction mode such
as transmission or reflection (see for example, Fig.8.19). One of the
disadvantages of a SSD is the so-called pulse pile-up problem, when a
large amount of photons are available in a very short time. The use of
a pulse pile-up rejector, the counting rate below about 10^4 per second
and the dead-time correction are taken into account for reducing this

trouble.

With this careful examination in mind, the new anomalous x-ray scattering measurement has recently been carried out at the A2 beam station of Cornell High Energy Synchrotron Source (CHESS) Laboratory (Ithaca, New York, U.S.A.). The experimental set-up for this measurement is schematically given in **Fig.8.20.** The intense white radiation from the storage ring was monochromatized by a channel-cut Si (200) crystal with the energy resolution of 14 eV at 20 keV. The sample of a $Mo_{50}Ni_{50}$ glass produced by the high rate sputtering deposition was mounted on a Picker four axis diffractometer.

In order to reduce the air scattering intensity, the sample was maintained in He gas atmosphere. The x-ray scattering intensity was measured by a intrinsic Ge SSD. The incident beam intensity was monitored by measuring the Compton scattering intensity from a Kapton (polyimide) tape placed in the incident beam path by using another SSD. The incident beam intensity fluctuates for several reasons such as the natural decay of the ring current in the storage ring. However, the fluctuation in the incident beam intensity was minimized by this experimental set-up and then the stability and reproducibility of measured intensity data were satisfactorily ensured.

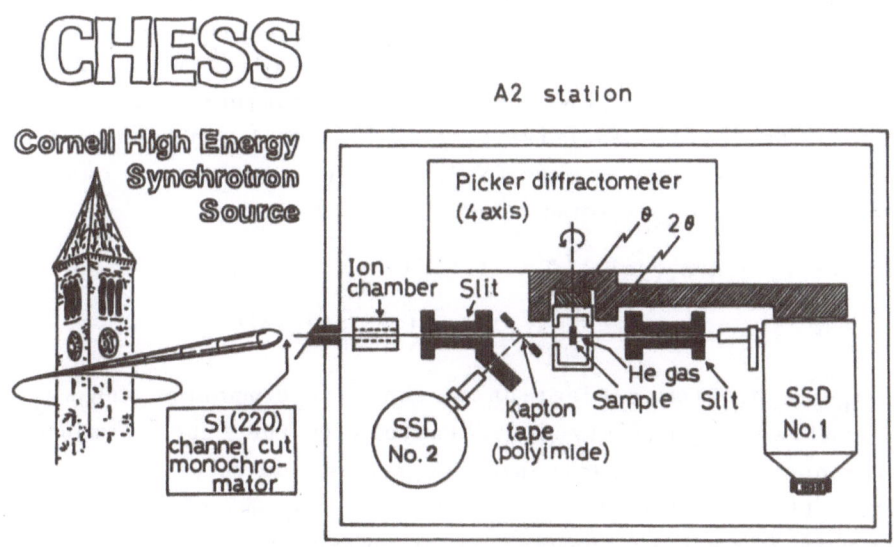

Fig.8.20 Schematic diagram of the apparatus for the anomalous x-ray scattering experiment used at CHESS.

The measurement with several energies near the absorption edge carried out at CHESS has been concerned with the energy dependence of x-ray scattering intensity arising from the difference in the anomalous dispersion effect. The basic concept of this energy derivative technique slightly differs from that of the direct anomalous x-ray scattering technique reported previously, although the principle of the anomalous x-ray scattering is unchanged.

As already suggested by Shevchik (1977) and Munro (1982), the energy derivative technique is found to be about an order of magnitude better than the direct anomalous x-ray scattering technique and it also appears to be suitable for multi-component disordered materials such as ternary systems. This energy derivative technique of the anomalous x-ray scattering was first used by Fuoss et al. (1981) with the synchrotron radiation at Stanford under the name of the differential anomalous scattering (DAS) method and gave the information that the threefold-coordinated model (Bienenstock 1973) is more favorable for describing the structural features of the Ge-Se glasses, compared with the fourfold-twofold coordinated model (Betts et al. 1970). Although the experimental details have not been reported yet in their paper (Fuoss et al. 1981), their conclusion could not be obtained by the conventional diffraction methods.

In order to facilitate the understanding of this energy derivative technique of the anomalous x-ray scattering, the schematic diagram for the experimental mode is provided in **Fig.8.21.** The essential equations are also given below for convenience.

The structurally sensitive part of the total coherent x-ray scattering intensity for a binary (A-B) disordered material can be described by the following form;

$$I(q,E) = c_A^2 f_A^2(q,E) a_{AA}(q) + c_B^2 f_B^2(q,E) a_{BB}(q)$$
$$+ 2c_A c_B f_A(q,E) f_B(q,E) a_{AB}(q) \qquad (8.10)$$

where c_i is the atomic fraction of the i-th component and $a_{ij}(q)$ is the so-called partial structure factors in the Faber-Ziman's form. When the energy of the incident x-rays is tuned to close vicinity of the absorption edge of the A-component, (for example, the energies of a and c in Fig.8.21), the variation detected in intensity may be attributed **only to the change** in the atomic scattering factor $f_A(q,E)$. Thus, one can obtain the following equation;

$$\Delta I_A(q) \simeq [\frac{\partial I(q,E)}{\partial E}]_q \propto 2c_A^2 f_A^2(q,E)a_{AA}(q) + 2c_A c_B f_B(q,E)a_{AB}(Q) \qquad (8.11)$$

Hence, the quantity of $\Delta I_A(q)$ or $[\partial I(q,E)/\partial E]_q$ associated with the A-component contains information of **two partial structure factors**, $a_{AA}(q)$ and $a_{AB}(q)$ and its Fourier transform obtained by eq.(8.12) provides the **average distribution function around an A atom** without complete separation of partial functions.

$$\rho_A(r) = \frac{1}{2\pi^2 \rho_o r} \int_0^\infty q\,\Delta I_A(q)\,\sin(q \cdot r)\,dq \qquad (8.12)$$

In other words, the **environmental structure around a desired atom** can be evaluated from the measurement using the energy derivative technique of the anomalous x-ray scattering. In a sense, this information is found to be very similar to the EXAFS data, although the anomalous x-ray scattering gives the long-range atomic correlation.

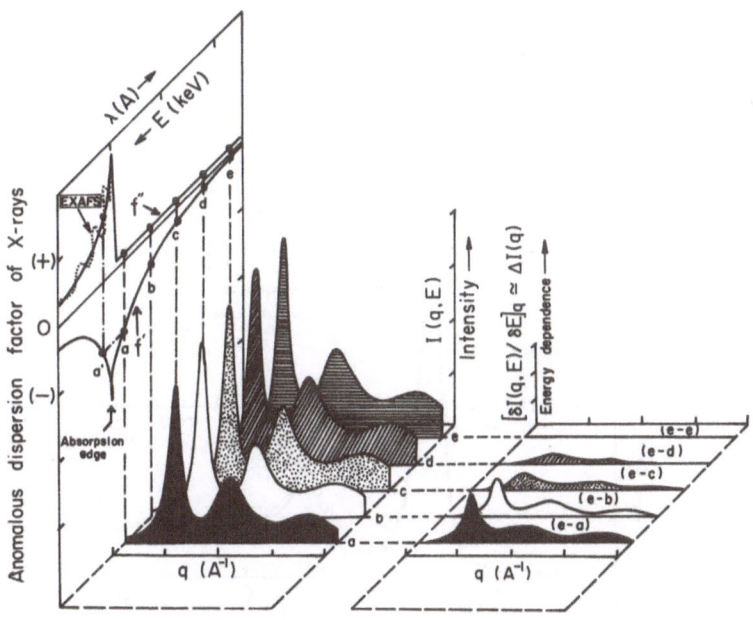

Fig.8.21 Schematic diagram of the experimental mode for disordered materials by the anomalous x-ray scattering.

In order to prevent strong fluorescent radiation from sample, the energies below the absorption edge are usually employed for the anomalous x-ray scattering measurement and in such an energy region, the change in the real part of the anomalous dispersion f' is dominant, because the imaginary part f" is almost constant and small as easily seen in Fig.8.21. Along the line of such experimental mode, the x-ray scattering intensity for the $Mo_{50}Ni_{50}$ glass was measured at seven energies (19.309, 19.478, 19.650, 19.781, 19.870, 19.914 and 19.959 keV) near the K-absorption edge (20.004 keV) of Mo atoms in transmission geometry using the experimental set-up of Fig.8.20 at the Cornell University Synchrotron Radiation Laboratory (CHESS).

The measured intensity significantly depends upon the incident x-ray energy, as shown in **Fig.8.22** and this observation should be attributed to the variation in the real component of the Mo anomalous

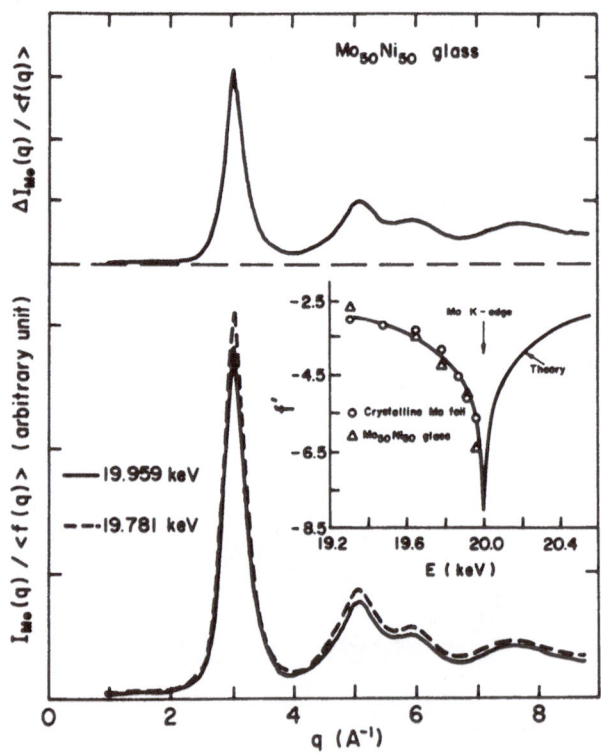

Fig.8.22 Energy dependence of x-ray scattering intensity of the $Mo_{50}Ni_{50}$ glass measured at energies near the K edge of Mo atoms (E_K=20.004 keV). The values of f' are -4.57 for E=19.781 keV and -6.18 for E=19.959 keV, respectively.

dispersion factor of f' where f' = -4.57 for E = 19.781 keV and f' = -6.18 for E = 19.959 keV, respectively. These values were determined from the measurement of both Mo foil and $Mo_{50}Ni_{50}$ glass and agree well with theoretical values by the Cromer-Liberman's scheme, as shown in Fig.8.22.

Further analysis including the correction for absorption, air scattering, multiple scattering and Compton scattering has been made with respect to the experimental data obtained at seven energies and then the total structure factor $a_T(q)$ and the environmental partial structure factor around Mo atom $a_{Mo}(q)$ were determined. The definition of these functions are given as follows;

$$\left.\begin{aligned} a_T(q) &= I_T(q)/\langle f(q)\rangle^2 \\ I_T(q) &= c_A^2\bar{f}_A^2(q)a_{AA}(q)+c_B^2\bar{f}_B^2(q)a_{BB}(q)+2c_Ac_B\bar{\bar{f}}_A(q)\bar{f}_B(q)a_{AB}(q) \end{aligned}\right\} \quad (8.13)$$

$$\left.\begin{aligned} a_T(q) &= I_T(q)/\langle f(q)\rangle^2 \\ I_A(q) &= 2c_A\bar{f}_A(q)a_{AA}(q) + 2c_Ac_B\bar{f}_B(q)a_{AB}(q) \end{aligned}\right\} \quad (8.14)$$

$$\langle f(q)\rangle = c_A\bar{f}_A(q) + c_B\bar{\bar{f}}_B(q) \quad (8.15)$$

$$f_i(q,E) = \bar{f}_i(q) + \Delta f_i'(E) \qquad \sum_n \Delta f_i'(E_n) = 0 \quad (8.16)$$

where the sum over n is taken for the number of energies employed for the measurement (n = 7 in the present case). Here, the difference in the q-dependence between $f_A(q)$ and $f_B(q)$ is assumed to be negligible, similar to the Warren-Krutter-Morningstar approximation (see for example, Wagner 1978). The following relation is also worthy of note.

$$2a_T(q) = \frac{\bar{f}_A(q)}{\langle f(q)\rangle} a_A(q) + \frac{\bar{f}_B(q)}{\langle f(q)\rangle} a_B(q) \quad (8.17)$$

This equation implies that knowledge of $a_T(q)$ and $a_A(q)$ allows the evaluation of $a_B(q)$. This is one way to obtain the environmental partial structure factor $a_B(q)$, at least, in cases when the measurement at energies near the absorption edge of the B-component cannot be assessed yet.

The structure factors $a_T(q)$ and $a_{Mo}(q)$ determined by this data processing are shown in **Fig.8.23** without any artificial process such as smoothing of data. Compared with the earlier effort by the anomalous x-ray scattering using the characteristic Kα radiations (see for example, Waseda 1981), the statistical accuracy is significantly improved by using the synchrotron radiation source. The present results provide the information in a wider q range with sufficient accuracy that the so-called truncation error is not significant in Fourier transformation of eq.(8.12) for deriving the environmental radial distribution function.

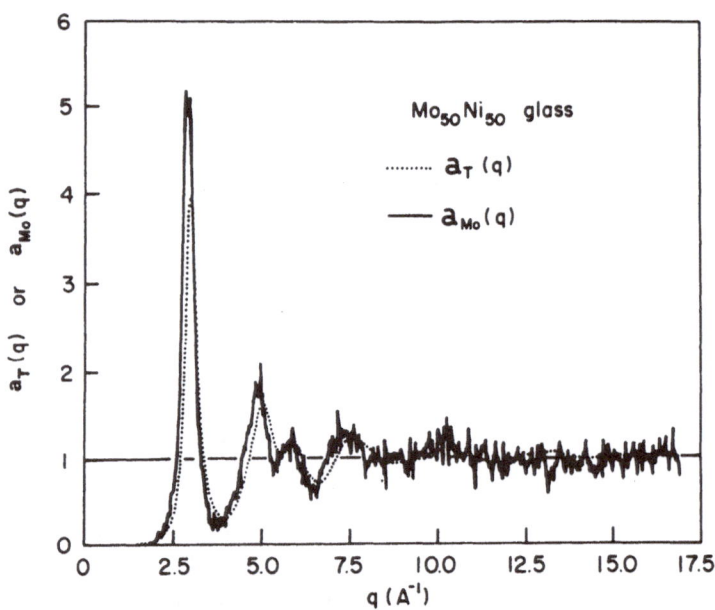

Fig.8.23 Total and environmental partial structure factors of the $Mo_{50}Ni_{50}$ glass determined by the anomalous x-ray scattering with a synchrotron radiation source (Aur et al. 1983a).

The resultant RDFs shown in **Fig.8.24** clearly indicate that the local environmental structures including the second nearest neighbors around Mo or Ni atoms are quite distinct from each other. For example, the first peak around Mo atom is found to be broader than that around

Ni atom, although the difference in peak positions should be attributed to the atomic size difference. More detailed information regarding the structure for the $Mo_{50}Ni_{50}$ glass has been reported by Aur et al.(1983a) together with the model calculation and needs no repeat description here. Nevertheless, it is the author's intention to suggest the usefulness of the anomalous x-ray scattering for determining the fine structure of multi-component disordered materials, when coupled with the synchrotron radiation source. We may also add that a similar structural investigation by applying the anomalous x-ray scattering with a synchrotron radiation source (Stanford) has recently been carried out for determining the structure of the Mo-Ge glasses by Kortright and Bienenstock (1984) and only the resultant RDFs are given in **Fig.8.25.**

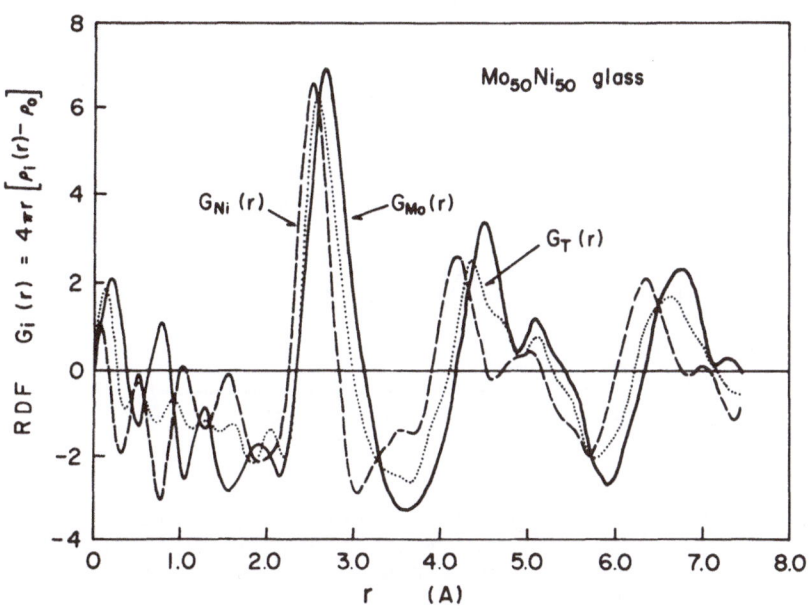

Fig.8.24 Total and environmental partial distribution functions of the $Mo_{50}Ni_{50}$ glass determined by the anomalous x-ray scattering with a synchrotron radiation source (Aur et al. 1983a).

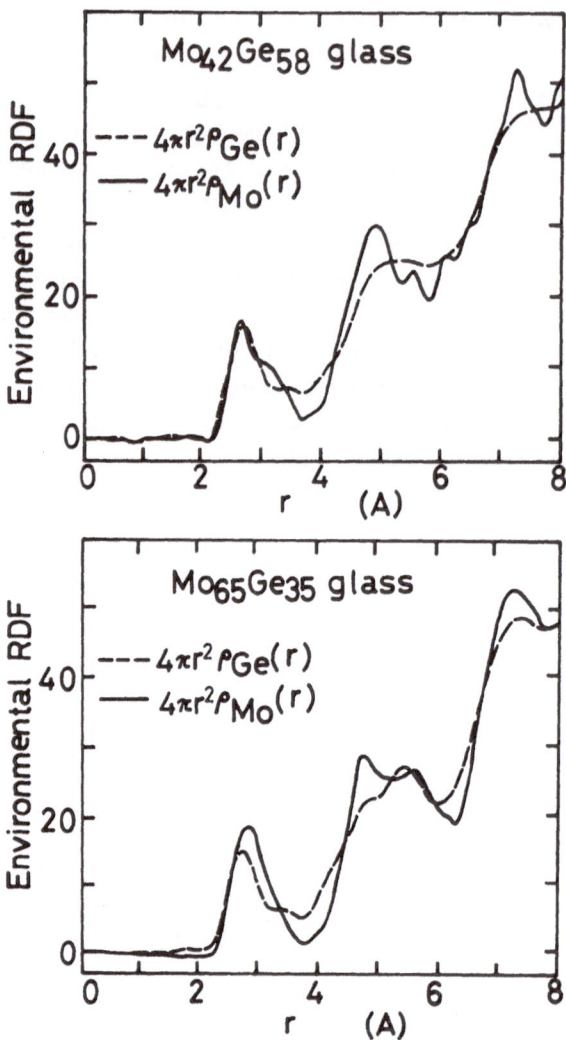

Fig.8.25 Environmental partial distribution functions of Mo-Ge
glasses determined by the anomalous x-ray scattering
with a synchrotron radiation source (Kortright and
Bienenstock 1984).

Considering all factors mentioned in this article with some
selected examples, it may be concluded that the **anomalous x-ray scattering technique** will become, in the near future, **one of the most
powerful and reliable tools** for structural characterization of disordered materials, in parallel with recent progress of the synchrotron
radiation source.

CHAPTER 9
RELATIVE MERITS OF ANOMALOUS X-RAY SCATTERING AND ITS FUTURE PROSPECTS

A number of techniques of x-rays, neutrons and others have been employed to determine the structure of disordered materials such as liquids, amorphous solids and solid fast ion conductors and each technique has, of course, its own advantages and disadvantages. The relative merits of these various techniques for structural characterization of disordered materials have already been provided in detail in some specialized monographs (see for example, Temperly et al. 1968, Beer 1972, Lee and Teo 1981, Guntherodt and Beck 1981) and the intention of this article is not to duplicate their description. We give here some essential points of the anomalous x-ray scattering with an emphasis on future developments.

As is well-known and also described in Appendix 2, the EXAFS technique is undoubtedly one of the most powerful methods for determining the local environmental structure of a specific atom in disordered materials. However, the anomalous x-ray scattering technique is much more straightforward, at least, theoretically. For example, the theoretical difficulties associated with the electron phase shifts and mean free path still make it impossible to obtain reliable information from the EXAFS measurement alone, particularly for systems with unknown structures like disordered materials. This point has already been stressed (see for example, Lee et al. 1981). In a sense, the power of the EXAFS may be somewhat over-emphasized in the last few years, although it shows several impressive advantages mentioned previously (see also Appendix 2). On the other hand, the anomalous x-ray scattering does not require the information of phase shifts and mean free path. The following disadvantages of the EXAFS measurement may be also noted:

(1) In a binary disordered alloy, when there are only two absorption edges available, it is not possible to evaluate unambiguously the three partial functions by the EXAFS measurements alone.

(2) The EXAFS measurement usually provides accurate information on the nearest neighbor correlations for a specific atom (the distance and its coordination number), it is ,however, relatively difficult to obtain information on higher ordering such as the second and third

nearest neighbor correlations, mainly due to the lack of low wave vector (q) information including the q-range where the usual first peak appears in the structure factor.

It should be noted that the structure of metallic glasses is known to primarily characterized by the second peak splitting in the RDF and we could not distinguish the structural difference between the glassy state and the liquid state for metallic disordered system by using only information of the nearest neighbor correlations. The anomalous x-ray scattering technique reduces these difficulties by making available three partial functions of RDFs.

The EXAFS spectrum is known to correspond to the energy dependence of the imaginary component f" of the anomalous dispersion factors as described in Chapter 5-7 (see also Fig.8.21). On the other hand, the intensity difference obtained from the particular measurement in the anomalous x-ray scattering with two different energies where the imaginary component f" is nearly constant and small (for example, the energies a and b in Fig.8.21 or λ_2 and λ_3 in Fig.5.1) provides information about only atoms scattering x-ray anomalously. In other words, this intensity difference should be attributed to the change due to only the real component f' of the specific atoms. For these reasons, the experimental data obtained by the energy derivative technique in the anomalous x-ray scattering (or the differential anomalous scattering named by Fuoss et al. 1981) is, in a sense, very similar to the information given by the EXAFS, although there are differences in detail as mentioned previously. In order to facilitate the understanding of the relationships between the EXAFS and the anomalous x-ray scattering, the schematic diagram is given in **Fig.9.1** with respect to the structural investigation for disordered materials.

One of the advantages of the anomalous x-ray scattering over the EXAFS method is that the environmental partial structure factor $a_i(q)$ can be determined in a wide range of wave vector q, including the first peak of $a_i(q)$ which is often missed by the EXAFS method, because of the cut-off (usually 3-5 A^{-1}) in the low wave vector region (see for example, Lee and Teo 1981). In other words, only the anomalous x-ray scattering gives the local environmental structure including not only 1st but also 2nd and 3rd nearest neighbor atomic correlations as a function of distance. This contrasts with the EXAFS results. **Figure 9.2** provides such information, where the EXAFS peaks are corrected for the phase shifts. As is generally the case, the first peak of the EXAFS slightly shifts toward a shorter distance by about 0.1 A (Teo et al. 1983). However, the present author believes that the overall agreement in the nearest neighbor correlations is rather well recognized.

It may be worth mentioning for disordered materials that, the anomalous x-ray scattering gives better coordination numbers, whereas the EXAFS provides direct information regarding the nature of the atomic pair for the peaks (see for example, Lee and Teo 1981). Therefore, the anomalous x-ray scattering data could supplement the EXAFS data or <u>vice versa</u> and the EXAFS method is superior to the anomalous x-ray scattering is in some special cases such as for study of the environmental structure of minor constituent elements in the desired samples.

As mentioned previously, the anomalous x-ray scattering can be applied to the structural study of various systems with only a few exceptions such as light elements, when coupled with the synchrotron

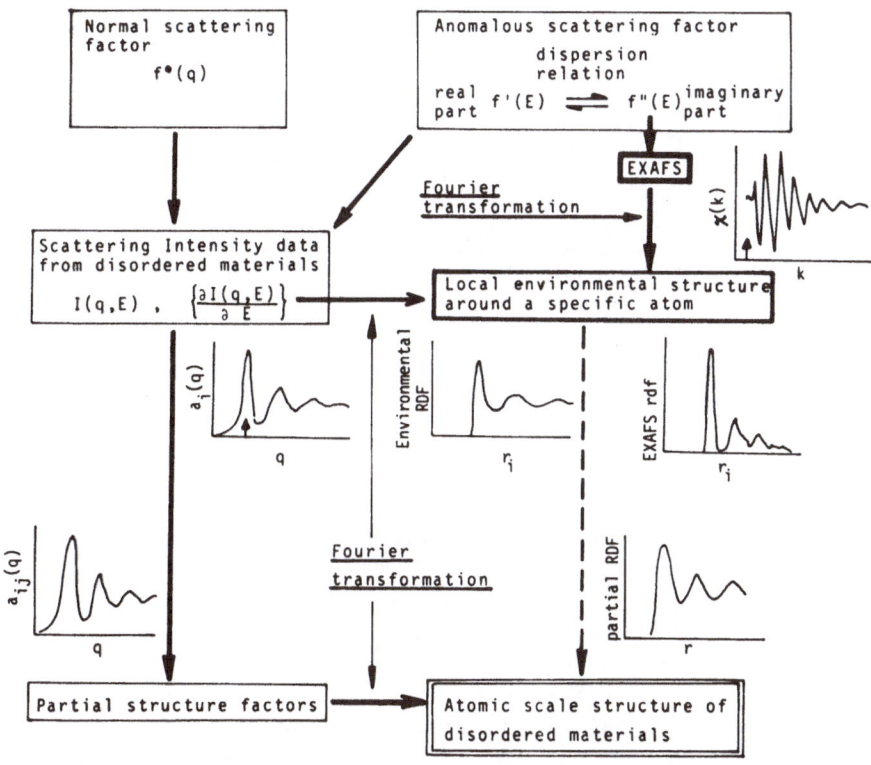

Fig.9.1 Schematic diagram of the relationships between the EXAFS method and the anomalous x-ray scattering for structural characterization of disordered materials.

radiation source by making available good continuous spectrum over a much wider energy range. This is also an advantage compared with the anomalous scattering and isotope substitution techniques with neutrons, because of the following reasons. The anomalous neutron scattering is very limited to rather rare elements such as ^6Li, ^{10}B, ^{113}Cd, ^{149}Sm, ^{153}Eu and ^{157}Gd, although the change in the anomalous dispersion factors of neutrons is several times larger than the x-ray case as exemplified by **Fig.9.3** using the results of ^{113}Cd (Peterson and Smith 1961). The isotope substitution technique is also limited to the number of appropriate isotopes and this method is known to be sometimes affected by the fact that some structural behavior depends upon the isotope (see for example, Bosio et al. 1981).

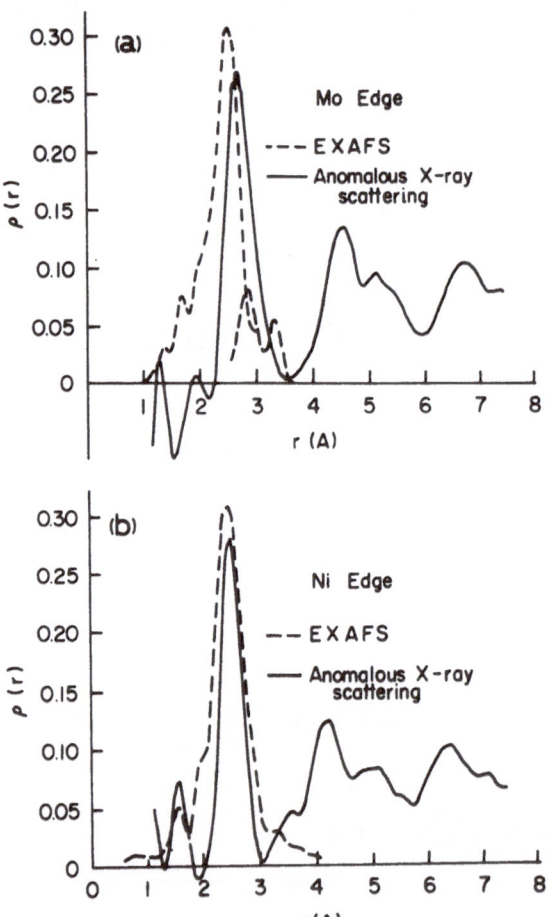

Fig.9.2 Environmental structural information around a Mo atom (a) or around a Ni atom (b) for the $Mo_{50}Ni_{50}$ glass. Solid lines: anomalous x-ray scattering (Aur et al. 1983a), broken lines: EXAFS method (Teo et al. 1983).

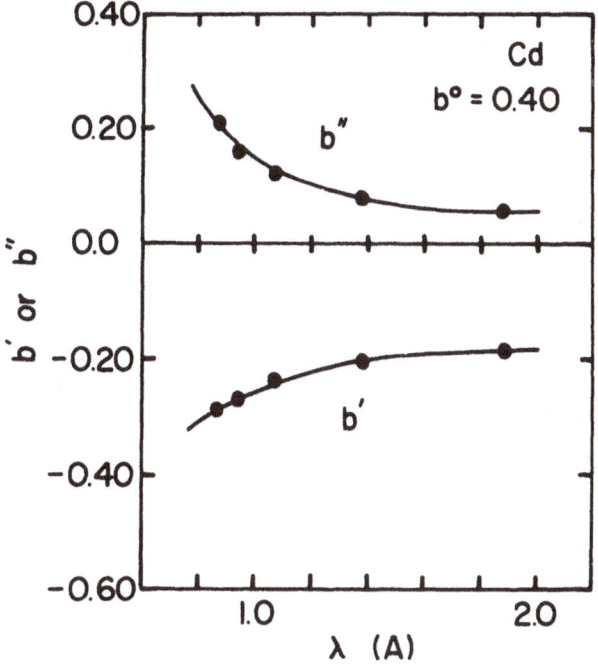

Fig.9.3 Anomalous dispersion factors b' and b" of Cd atoms
for neutrons (Peterson and Smith 1961).

As clearly demonstrated by the recent results of the $Mo_{50}Ni_{50}$
glass given in Chapter 8.3, the intensity difference or the energy
dependence of x-ray scattering intensity at constant q, obtained from
the experiments with different energies near the absorption edge of a
specific constituent, provides the environmental structure around only
atoms scattering x-ray anomalously, without complete separation of
partial functions. The characteristic absorption edge of various
elements are separated, at least, several hundred of eV, so that the
data collection by the energy derivative technique of the anomalous x-
ray scattering is feasible within the energy change of about 100-200
eV near the absorption edge with sufficient atomic sensitivity. In
this regard, the following extended method for structural study of
solutions may also be suggested.

The structural investigation of solutions has been recognized as
one of the most important research subjects in both aqueous solution
chemistry and hydrometallurgy. Such importance is again emphasized in
parallel with recent progress in modern biochemistry. One of the
important findings for these studies is the average distribution of

water molecules around specific ions which are described by the partial RDF in solutions. This is frequently characterized by the concept of **hydration numbers** in solutions. The use of the anomalous x-ray scattering will again bring about a significant breakthrough in this research subject. The principles and characteristic features of this new method for separating the partial structures related to the hydration numbers in solutions are given below using the MX_n type electrolytic solution as an example. This may be convenient for the future development in the anaomlous x-ray scattering technique.

The structural sensitive part $F(q)$ of measured x-ray scattering intensity for the MX_n solution (MX_n-H_2O system) may be expressed by the following ten partial structure factors, $a_{OO}(q)$, $a_{HH}(q)$, $a_{OH}(q)$, $a_{MM}(q)$, $a_{XX}(q)$, $a_{MX}(q)$, $a_{XO}(q)$, $a_{MO}(q)$, $a_{MH}(q)$ and $a_{XH}(q)$, as already suggested by Soper et al (1977). However, when the energy of the incident x-rays is tuned to close vicinity of the absorption edge of the component M, the observed variation in intensity $F_M(q)$ contains information of four partial structure factors, $a_{MM}(q)$, $a_{MX}(q)$, $a_{MO}(q)$ and $a_{MH}(q)$ and may be given by the following form in a manner similar to the first-order difference scattering of neutrons with isotope (see for example, Soper et al. 1977, Enderby and Neilson 1981).

$$\Delta F_M(q) = c_M^2 (f_M^2 - f_M^{\#2})[a_{MM}(q)-1] + 2nc_M^2 f_x (f_M - f_M^{\#})[a_{MX}(q)-1]$$

$$+ 2c_O c_M f_O (f_M - f_M^{\#})[a_{MO}(q)-1] + 4c_O c_M f_M (f_M - f_M^{\#})[a_{MH}(q)-1] \quad (9.1)$$

$$= \alpha[a_{MM}(q)-1] + \beta[a_{MX}(q)-1] + \gamma[a_{MO}(q)-1] + \delta[a_{MH}(q)-1] \quad (9.2)$$

$$= \alpha a_{MM}(q) + \beta a_{MX}(q) + \gamma a_{MO}(q) + \delta a_{MH}(q) - (\alpha+\beta+\gamma+\delta) \quad (9.3)$$

where c_O and c_M are the atomic fraction of oxygen and cation, respectively, f_i is the usual atomic scattering factor of the component i excluding the q-dependence for simplification, whereas f_M denotes the atomic scattering factor at an energy close to the absorption edge of M species.

In solutions, the atomic fraction of solute, c_M in the present case, is generally small and therefore the coefficients α and β are considered to be small compared with the coefficients γ and δ. For this reason, the following simplification can be made.

$$\Delta F_M(q) + (\alpha+\beta+\gamma+\delta) \cong \gamma a_{MO}(q) + \delta a_{MH}(q) \rightarrow \gamma a_{MO}(q) \qquad (9.4)$$

In x-ray diffraction, the hydrogen atoms have no strong scattering ability and thus the quantity $\Delta F_M(q)$ of measured intensity data may be reasonably approximated by the partial structure factor of $a_{MO}(q)$ of which Fourier transform in the usual manner (see eq.(8.12)) gives the direct information on the distribution of oxygen (water molecules in solutions) around M species in solutions presently considered.

The present new technique by applying the anomalous dispersion effect of x-rays may be superior to the isotope substitution method with neutrons where the structure is automatically assumed to maintain identical upon substitution by the isotope, because such an assumption strongly depends on the systems and is frequently questionable, even if the isotopes are chemically similar and their size difference is small. For example, the low angle behavior of water (H_2O) has been found to be slightly different from that of heavy water (D_2O) as suggested by Bosio et al. (1981).

The anomalous x-ray scattering proposed here may provide the answer to the question that the hydration numbers in $NiCl_2$ electrolytic solutions determined by the EXAFS measurements (Sandstrom 1979) differ from the values evaluated by the isotope substitution method of neutrons (Soper et al. 1977). Of course, a number of preparations will also be required, before this new technique is recognized as a reliable tool for structural characterization of solutions (see for example, Waseda and Sakuma 1982).

The anomalous x-ray scattering technique for determining the fine structure of disordered materials has not yet been completed. Unless a reasonably high energy absorption edge is used, the disadvantage due to the limited q-range available (q_{max} value less than about 7 A^{-1}) is unavoidable and in such a case, a careful interpretation of the resultant RDF is required. Nevertheless, as mentioned in this article with some selected examples, we have already built up a rather wider base for the anomalous x-ray scattering technique including the present status and future directions on its application to structural study of various materials. A coupled angular scanning mode using the energy sensitive solid state detector and the synchrotron radiation source unanimaously appears to hold promise that the anomalous (resonance) x-ray scattering technique should basically work very well and its potential power, in the present author's view, may not be overemphasized. When this technique is completed, it should be possible to provide a significant impact on the relationships between atomic scale structure and various properties of multi-component disordered materials with much higher reliability than the EXAFS measurement.

We may also add that the anomalous x-ray scattering technique would be valuable not only for the structural study of disordered materials but also of other materials in a variety of states. For example, high polymers, one-dimensional liquids such as $Hg_{3-\delta}AsF_6$, intercalated graphite, or some specialized topics such as cation distribution in spinels, short range order in crystalline alloys (see Chapter 8.1) and macromolecules or macromolecular complexes particulaly in biological medium (see for example, Blasie and Stamatoff 1981). If further investment with a highest priority could be continued in developing the use of the anomalous dispersion phenomena of x-rays with the synchrotron radiation source, the anomalous (resonance) x-ray scattering will achieve the goal in a relatively short period.

APPENDIX 1

Local Structural Fluctuations and the Long Wavelength Limit of the Structure Factors in Disordered Systems

In order to facilitate the fuller understanding of the concept and usefulness for the Bhatia-Thornton partial structure factors, some essential points are given in this appendix.

Consider the local fluctuation in the number density for a subsystem with a small volume v containing N atoms and then the fluctuation of number density $\Delta n(r)$ is expressed by;

$$\Delta n(\vec{r}) = n(\vec{r}) - \bar{n} = \sum_j \delta(r-r_j) - \bar{n} \tag{A1-1}$$

where $\bar{n} = N_0/V_0$, N_0 and V_0 are the number of atoms and the volume in the system, respectively. ($v \ll V_0$ and $N \ll N_0$). By using the Fourier transformation, the following equations can be made:

$$\Delta n(\vec{q}) = \int n(\vec{r}) \exp(-i\vec{q}\cdot\vec{r}) \, d\vec{r} = \sum_j \exp(-i\vec{q}\cdot\vec{r}) - \bar{N}\delta_{\vec{q},0} \tag{A1-2}$$

$$|\Delta n(\vec{q})|^2 = \sum_{jk}\sum \exp\{-i\vec{q}\cdot(\vec{r}_j-\vec{r}_k)\} - 2N\bar{N}\delta_{\vec{q},0} + \bar{N}^2\delta_{\vec{q},0} \tag{A1-3}$$

Taking the limiting case of $\vec{q} \to 0$ in eq.(A1-3), we obtain;

$$\lim_{\vec{q}\to 0}|\Delta n(\vec{q})|^2 = N^2 - 2N\bar{N} + \bar{N}^2 = (N - \bar{N})^2 = (\Delta N)^2 \tag{A1-4}$$

The structure factor of a subsytem may be given in the following equation using the Fourier transform of the number density fluctuation;

$$S(\vec{q}) = \bar{N}^{-1}<|\Delta n(\vec{q})|^2> = N^{-1}<\sum_{jk}\sum \exp\{-i\vec{q}\cdot(\vec{r}_j-\vec{r}_k)\}> - \bar{N}\delta_{\vec{q},0} \tag{A1-5}$$

Thus, the following well-known relation (see for example, Landau and Lifschitz 1958) can be obtained for the long-wavelength limit of a one-component disordered system;

$$S(0) = \bar{N}^{-1}<(\Delta N)^2> \tag{A1-6}$$

Let consider, next, a subsystem containing two kinds of atoms 1 and 2 in a similar way. When there are N_{0j} of j-type atoms (j = 1 and

2) in the volume V_0 including a total number of atoms N_0, the average values of j-type atoms and the total number density, $\bar{n}_j$ and $\bar{n}$ are given as follows;

$$\bar{n}_j = n_{oj}/V_0 = \bar{N}_j/v$$

$$\bar{n} = N_0/V_0 = \bar{N}/v \tag{A1-7}$$

Then, the number density is expressed by;

$$n(\vec{r}) = n_1(\vec{r}) + n_2(\vec{r}) \tag{A1-8}$$

where

$$n_j(\vec{r}) = \sum_{k=1}^{N_j} (\vec{r} - \vec{r}_{jk}) \tag{A1-9}$$

Hence, the deviation of the number density from the average number density $\Delta n_j(\vec{r})$ and its Fourier transform $N_j(\vec{q})$ can be written by;

$$\Delta n_j(\vec{r}) = n_j(\vec{r}) - \bar{n} \tag{A1-10}$$

$$N_j(\vec{q}) = \int \exp(-i\vec{q}\cdot\vec{r}) \Delta n_j(\vec{r}) d\vec{r} = \sum_{k=1}^{N_j} \exp(-i\vec{q}\cdot\vec{r}_{jk}) - \bar{N}_j \delta_{\vec{q},o} \tag{A1-11}$$

By applying a similar procedure, we can define the Fourier transform $N(\vec{q})$ of the local deviation $\Delta n(\vec{r})$ in the total number density from its average value by the following equation.

$$N(\vec{q}) = N_1(\vec{q}) + N_2(\vec{q}) = \sum_{k=1}^{N} \exp(-i\vec{q}\cdot\vec{r}_k) - \bar{N}\delta_{\vec{q},o} \tag{A1-12}$$

In the long-wavelength limit $(\vec{q} \to 0)$, $N(0) = N - \bar{N} \neq 0$ and then the value of $N(\vec{q} \to 0)$ given by eq.(A1-12) corresponds to the deviation of atomic number in a volume v from the average value of N. With respect to the quantity $N_j(\vec{q})$, a similar comment also holds.

On the other hand, the following relation can be obtained regarding the local fluctuation in the concentration c, in a way similar to the previous description.

$$c(\vec{r}) = c(\vec{r}) - \bar{c} = (V/\bar{N}) \left[(1-c)\Delta n_1(\vec{r}) - c \, \Delta r_2(\vec{r}) \right] \tag{A1-13}$$

$$C(\vec{q}) = \bar{N}^{-1} \left[(1-c)N_1(\vec{q}) - cN_2(\vec{q}) \right] \tag{A1-14}$$

where c $(= N_{01}/N_0 = N_1/N)$ denotes the avarage concentration of 1-type atoms in a volume v. When $\Delta n_1(r)$ and $\Delta n_2(r)$ vary with their respective average concentration c and (1-c), eq.(A1-13) implies the relation of $c(\vec{r})$ should be equal to zero. On the basis of eqs.(A1-12) and (A1-14), the following relation can readily be obtained.

$$N_1(\vec{q}) = cN(\vec{q}) + \bar{N}C(\vec{q})$$

$$N_2(\vec{q}) = (1-c)N(\vec{q}) - \bar{N}C(\vec{q})$$

$$\left.\right\} \qquad (A1-15)$$

Thus, we can define the scattering amplitude $A(\vec{q})$, attributed to the local fluctuation of density and concentration in a binary disordered system, as follows.

$$A(\vec{q}) = f_1 N_1(\vec{q}) + f_2 N_2(\vec{q}) = <f>N(\vec{q}) + N(f_1-f_2) C(\vec{q}) \qquad (A1-16)$$

where $<f> = c_1 f_1 + c_2 f_2$ and the q-dependence of the atomic scattering factor $f_i(q)$ is excluded here for simplification. Based on these expressions, the coherent x-ray scattering intensity can be obtained in the following form (Bhatia and Thornton 1970).

$$I_a^{coh}(q) = <f>^2 <N*(\vec{q})N(\vec{q})> + (f_1-f_2)^2 \bar{N}^2 <C*(\vec{q})C(\vec{q})>$$

$$+ <f>(f_1-f_2)\bar{N}<N*(\vec{q})C(q) + C*(\vec{q})N(\vec{q})> \qquad (A1-17)$$

$$= \bar{N} \left[<f>^2 S_{NN}(q) + \Delta f^2 S_{CC}(q) + 2<f>\Delta f S_{NC}(q) \right] \qquad (A1-18)$$

where $\Delta f = f_1 - f_2$ and $S_{NN}(q)$, $S_{CC}(q)$ and $S_{NC}(q)$ are the Bhatia-Thornton partial structure factors. Hence, we can immediately write the following usual formulas given in Chapter 3.

$$I_a^{coh}(q) = <f>^2 S_{NN}(q) + \Delta f^2 S_{NN}(q) + 2<f>\Delta f S_{NC}(q) \qquad (A1-19)$$

$$S_{BT}(q) = I_a^{coh}(q)/<f^2> = \frac{<f>^2}{<f^2>} S_{NN}(q) + \frac{<f>^2}{<f^2>} S_{CC}(q)$$

$$+ 2\frac{<f>\Delta f}{<f^2>} S_{NC}(q) \qquad (A1-20)$$

The Bhatia-Thornton partial structure factors give a useful physical significance in the long wavelength limit $(q \to 0)$. $S_{NN}(0)$ and $S_{CC}(0)$ correspond to the mean square fluctuation in the number density and the concentration, respectively. Whereas $S_{NC}(0)$ indicates the cross correlation between these two fluctuations. They are given by the following equations:

$$S_{NN}(0) = \langle(\Delta N)^2\rangle/\bar{N}$$

$$S_{CC}(0) = \bar{N}\langle(\Delta c)^2\rangle$$

$$S_{NC}(0) = \langle\Delta N \cdot \Delta c\rangle$$

(A1-21)

We shall briefly summarize the relation between the long wavelength limit of the structure factor and the thermodynamic quantities in disordered systems. This research subject is recently drawing attention, because of the introduction of $S_{CC}(0)$ value by making available a direct link between the experimental activity data with the structural information (see for example, Bhatia 1977, Tamaki et al. 1980).

According to the discussion in statistical thermodynamics (Landau and Lifschitz 1958), the number density fluctuation is connected with the isothermal compressibility χ_T in the following equation.

$$\langle(\Delta N)^2\rangle = \frac{\bar{N}^2}{V} k_B T \chi_T$$

(A1-22)

where k_B is the Boltzmann constant and T is the absolute temperature. Hence, we obtain quite readily the long wavelength limit of the structure factor $S(0)$ as follows.

$$S(0) = \rho_o k_B T \chi_T$$

(A1-23)

where ρ_o is the average number density. The partial structure factors in the Faber-Ziman and Ashcroft-Langreth forms are also related to the isothermal compressibility, which are analogous to a one-component system as follows.

$$\rho_o k_B T \chi_T = \frac{(c_1 a_{11}(0) + c_2)(c_2 a_{22}(0) + c_1) - c_1 c_2 \left[a_{12}(0)-1\right]^2}{1 + c_1 c_2 \left[a_{11}(0) + a_{22}(0) - 2a_{12}(0)\right]}$$

(A1-24)

$$\rho_o k_B T \chi_T = \frac{S_{11}(0)S_{22}(0) - S_{12}^2(0)}{c_1 S_{11}(0) + c_2 S_{22}(0) - 2(c_1 c_2)^{1/2} S_{12}(0)}$$

(A1-25)

According to the discussion of Bhatia and Thornton (1970), the fluctuations of number density and concentration and their cross correlation can be written as;

$$<(\Delta N)^2> = N\left[\rho_o k_B T\chi_T + Nk_B T\theta^2/(\partial^2 G/\partial c^2)_{P,T,N}\right] \qquad (A1-26)$$

$$<(\Delta c)^2> = \bar{N}^{-1}\left[Nk_B T/(\partial^2 G/\partial c^2)_{P,T,N}\right] \qquad (A1-27)$$

$$<\Delta N\Delta c> = -Nk_B T\theta/(\partial^2 G/\partial c^2)_{P,T,N} \qquad (A1-28)$$

$$\theta = \frac{1}{V}(\frac{\partial V}{\partial c})_{P,T,N} = \frac{v_1 - v_2}{c_1 v_1 + c_2 v_2} \qquad (A1-29)$$

where G is the Gibbs free energy and v_i is the partial molar volume of i-component. Combining eqs.(A1-26) (A1-28) with eq.(A1-21), the long wavelength limit of the Bhatia-Thornton partial structure factors are given by the following equations.

$$S_{NN}(0) = \rho_o k_B T\chi_T + \bar{N}k_B T\theta^2/(\partial^2 G/\partial c^2)_{P,T,N} = \rho_o k_B T\chi_T + \theta^2 S_{CC}(0) \qquad (A1-30)$$

$$S_{CC}(0) = \bar{N}k_B T/(\partial^2 G/\partial c^2)_{P,T,N} = (1-c_i)\left[(\partial\ln a_i/\partial c_i)_{T,P}\right]^{-1} \qquad (A1-31)$$

$$S_{NC}(0) = -\bar{N}k_B T/(\partial^2 G/\partial c^2)_{P,T,N} = -\theta S_{CC}(0) \qquad (A1-32)$$

where a_i is the thermodynamic activity of i-component. The long wavelength limit of the Bhatia-Thornton total structure factor can also be expressed by the following equation in terms of the volume , the isothermal compressibility χ_T and $S_{CC}(0)$.

$$S_{BT}(0) = \frac{<f(0)>^2}{<f^2(0)>}\left[\rho_o k_B T\chi_T + \{\theta - \frac{f_1(0)-f_2(0)}{<f(0)>}\}^2 S_{CC}(0)\right] \qquad (A1-33)$$

Once the value of $S_{CC}(0)$ is evaluated from the thermodynamic data or the scattering experiment for a zero alloy ($<f> = 0$) first demonstrated by Ruppersberg and Egger (1975) on liquid Li-Pb alloys, the calculation of $S_{NN}(0)$, $S_{NC}(0)$ and $S_{BT}(0)$ is straightforward as long as the volume and the isothermal compressibility data are available. However, the evaluation of $S_{CC}(0)$ by diffraction method is still found to be technically difficult, because of the coincidence of the forward scattering with the primary beam. We may suggest that the extrapolation method in the low q measurements using a narrow beam x-ray diffraction in the transmission mode (see for example, Huijben et al. 1977) is one way to reduce this difficulty by making available the $S_{BT}(0)$ data and thus enabling us the calculation of $S_{CC}(0)$ in the manner of eq.(A1-33).

The values of $S_{CC}(0)$ can also be estimated by several thermodynamic models instead of the measured activity data, because only the Gibbs free energy of mixing is required for the evaluation of $(\partial^2 G/\partial c^2)_{P,T,N}$ term in eq.(A1-31). The detailed discussion on the evaluation of $S_{CC}(0)$ by using various models, such as the regular solution and the conformal solution models, has already been given by Bhatia and his colleague (see for example, Bhatia et al. 1973, Bhatia 1977) and need not be repeated here.

APPENDIX 2
Relatively New Techniques for Structural Determination of Disordered Materials

A large number of experimental effort has been devoted to the development of new methods or the improvement of methods presently used for determining the fine structure of disordered materials and thus several significant technical advances have been made in the last five years. Most of these advances have already been described in detail by several specialized monographs or review articles (see for example, Suzuki 1976, Wong 1981 and Egami 1981b). However, we give here the essential points for these relatively new techniques for convenience of discussion with the anomalous x-ray scattering.

(A) Extended X-ray Absorption Fine Structure (EXAFS) Technique

The recent availability of a high intensity x-ray source (synchrotron radiation) and the Fourier transformation analysis proposed by Stern et al.(1975) have resulted in the establishment of the so-called EXAFS technique as a tool for structural analysis of disordered materials (see for example, Wong 1981, Hayes and Boyce 1982), although the EXAFS signal itself has been known for a long time. Thus the EXAFS measurement is undoubtedly one of the most powerful methods for determining the local coordinate structure in disordered materials.

The EXAFS signal, as shown in Fig.A2-1 using the results of Ni film as an example (Oyanagi and Hosoya 1980), refers to the oscillatory modulation of the x-ray absorption coefficient within a few hundred eV beyond an absorption edge of a constituent atom in a given material. The EXAFS oscillations from the monotonic terms in the absorption coefficient due to both K and L shells can now theoretically be interpreted by the effect arising from the interference of the outgoing photoejected electron with the backscattered one emerging from the near neighbor surrounding atoms. Thus, the frequency of the EXAFS signal mainly depends on the correlation distance between the central absorbing atom and the neighboring atom on the one hand, and the amplitude of the EXAFS signal is strongly affected by the number and the backscattering ability of the neighboring atoms on the other.

The normalized EXAFS signal function $\chi(k)$ as a function of photon energy beyond an absorption edge is normalized to the monotonic ab-

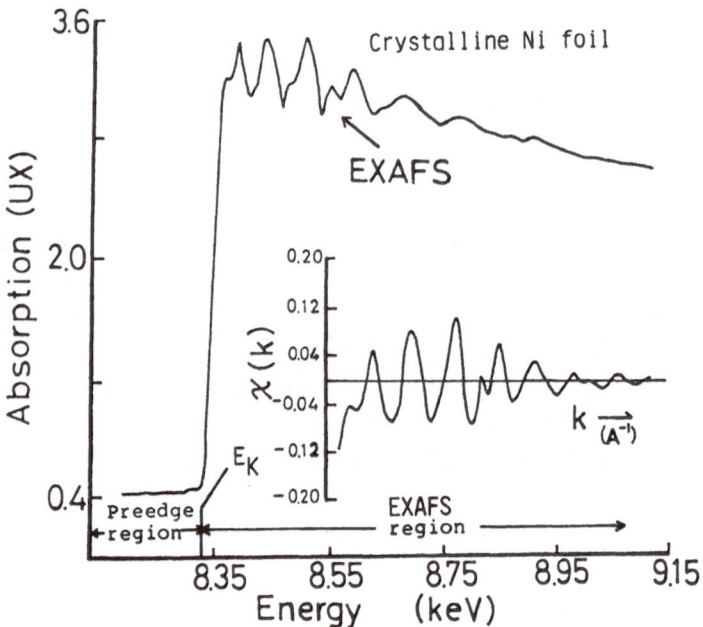

Fig.A2-1 Experimental spectrum of EXAFS for Ni film with the fine structural function $\chi(k)$ (Oyanagi and Hosoya 1980).

sorption term and generally expressed by the following form.

$$\chi(k) = \left[\mu(k) - \mu_o(k)\right]/\mu_o(k) \tag{A2-1}$$

where k corresponds to the wave vector of a photoelectron defined by $k = [2m(w-w_o)/h^2]^{1/2}$, w_o is the so-called inner potential and $\mu_o(k)$ is the absorption coefficient of the isolated atom indicating the smooth variation of k. Sayers et al.(1971) showed that a single scattering theory is adequate for interpreting the EXAFS signal under most circumstances and this was subsequently modified by Stern (1974) to a more generalized form as follows.

$$\chi(k) = -\frac{1}{k}\Sigma \frac{N_j}{r_j} t_j(2k)\exp(-2r_j/\lambda)\sin2(kr_j+\delta_j)\exp(-2k^2\sigma^2) \tag{A2-2}$$

where r_j and N_j are the radial distance and coordination number of the j-type atoms from an absorbing atom, respectively, $t_j(2k)$ corresponds to the backscattering matrix element encountered by the photoelectrons and λ is the mean free path of the photoelectrons. δ_j is the phase shift required to account for the potentials arising from both a central absorbing atom and neigboring atom (backscattering) and σ^2 is a Debye-Waller type factor corresponding to the root-mean-square positional fluctuations of the central and back scattering atoms.

By using the Fourier transformation, Sayers et al. (1971) proposed the fundamental equation between the radial distribution function (rdf of EXAFS) and the EXAFS signal in the following.

$$\phi_n(r) = (2\pi)^{-1/2} \int_{k_{min}}^{k_{max}} k^n \chi(k) \exp(-2ik \cdot r) dk \tag{A2-3}$$

We can immediately obtain the following equation for one simple example (n=1) from eq.(A2-3).

$$\phi_1(r) \propto \sum_j \frac{N_j}{r_j^2} \int_{k_{min}}^{k_{max}} t_j(2k) \sin\{2k(r-r_j)\} dk \tag{A2-4}$$

This leads to the delta function at the distance of $(r-r_j)$ and its weighting factor is proportional to the coordination number N_j in the limiting case of $k_{min} \to 0$ and $k_{max} \to \infty$. Hence, it is readily understood that the radial distribution function can be obtained from the EXAFS measurements. It should be noted, however, that the radial distribution function evaluated from the EXAFS signal (rdf) contains information about the distribution function of neighboring atoms for a central absorbing atom. On the other hand, the radial distribution function obtained by usual diffraction experiments (RDF) corresponds to the convolution of the EXAFS rdf, i.e., the average atomic distribution in the given system as a function of distance.

Figure 9.1 in chapter 9 gives the schematic representation for comparing x-ray diffraction with the EXAFS measurement for disordered materials. The EXAFS technique has the following particular advantages, compared with the conventional diffraction method of x-rays or neutrons.

(1) The EXAFS signal contains quantitatively accurate information about local environmental structure for an absorbing atom, even though its content is relatively small.

(2) The atomic sensitivity of the EXAFS signal is excellent and it is possible to separate the signals from materials including the elements whose atomic number is very close to each other, such as Fe and Ni, because their K-absorption edges are widely spread in energy

$(E_{K,Fe} = 7.112$ keV and $E_{K,Ni} = 8.332$ keV).

(3) The EXAFS technique, when coupled with a high intensity x-ray source of synchrotron radiation, enables one to follow the time dependent phenomena for a certain local structure.

These characteristic features clearly suggest that the EXAFS technique is very effective for revealing the compositional short range order (CSRO) in multi-component disordered materials. The EXAFS technique is also known to be applied for determining the environmental structure of minor component such as the local structure around a certain metallic ion in aqueous solutions where the presently available conventional diffraction methods are found to be almost useless.

Although the EXAFS technique shows many impressive advantages, it should be remembered that the theoretical difficulties make it impossible to obtain reliable information **from the EXAFS signal alone** at the present time, particularly **for highly disordered systems with unknown structures.** For example, the principle equation of (A2-2) includes only single scattering by the neighboring atoms, but it is obvious that the multiple scattering is not a negligible factor in some cases. The EXAFS signal is known to be strongly dependent upon the phase shifts, but their chemical and structural dependence are, both theoretically and experimentally, not completely revealed yet. Thus the results of several disordered materials obtained by the EXAFS technique (see for example, Wong 1981) are still to be regarded with some reservations (see for example, Cargill III 1982).

Other details concerning the EXAFS technique have been described elsewhere (see for example, Lee et al. 1981, Hayes and Boyce 1982); however, we may add the following points. As easily seen from Fig.8.21, the EXAFS signal corresponds to the energy dependence of the imaginary component f" of the x-ray anomalous dispersion terms, thus the anomalous x-ray scattering data could supplement the interpretation of the EXAFS data or _vice_ _versa_. For example, the anomalous x-ray scattering technique provides the desired information of the radial distance r_j and its coordination number N_j, for the so-called curve-fitting procedure so as to obtain a best fit between experimental and calculated EXAFS signal on a trial and error basis. As suggested in the previous review articles (see for example, Lee et al. 1981), the important information at small k (less than about 3.5 A^{-1}) is not employed in the EXAFS measurement, because of the restriction related to the near-edge phenomena and the single scattering theory (see also

Fig.9.1 or Fig.A2-1). Then, the normalized EXAFS signal function $\chi(k)$ is set equal to zero over such wave vector region in the Fourier transformation of eq.(A2-4). This substitution of ambiguous information for the missing data may frequently give rise to the EXAFS rdf with relatively large experimental uncertainty.

(B) Time-of-Flight (TOF) Pulsed Neutron Diffraction and Energy Dispersive X-ray Diffraction (EDXD)

One of the most important requirements in the structural analysis of disordered materials is to reveal the local ordering structure and its distribution and then the high resolution radial distribution function (RDF) will aid in solving this problem. Of course, the EXAFS technique, in principle, gives the answer on this requirement, but its reliability is still far from complete for disordered materials. An alternative method for obtaining the local ordering structure of disordered materials is to measure the structure factor at much higher values of q (wave vector), thereby giving greatly improved real space resolution.

For this purpose, both time-of-flight (TOF) pulsed neutron diffraction and energy dispersive x-ray diffraction (EDXD) have been employed and some significant progress, particularly in the application of these techniques to structural study of disordered materials, has been made recently (for example, Suzuki 1976, Egami 1981b). Since both techniques are based on variable energy measurements in which continuous neutron or x-ray sources are used, we can determine the structure factor in higher wave vector region, say up to 30 A^{-1}, which is beyond the usual limit of 17 A^{-1} for conventional x-ray and neutron diffractions. Thus, the high resolution RDF, particularly with higher resolution of the near neighbor region (about 2-5 A) can be obtained. A brief description regarding this characteristic feature is as follows.

The pulsed neutron produced by an electron LINAC of which energy spectrum indicates a relatively higher intensity in the epithermal region with a wavelength shorter than 0.5 A and thus allows TOF pulsed neutron diffraction to give the high wave vector (q) measurements by enabling us the use of a shorter wavelength region. On the other hand, the white radiation is used as an x-ray source in the EDXD and the diffracted photons are detected as a function of energy by an energy sensitive solid state detector. Therefore, the use of higher energies up to 50 keV (corresponding to 0.25 A) makes it possible to measure the structure factor in higher q region in the EDXD measurement.

Details of the instrumentation and data processing for these two techniques have been fully described (Sinclair et al. 1974, Suzuki 1976, Egami 1981b) and thus not duplicated here. However, the following characteristic features of the method of TOF pulsed neutron diffraction or EDXD may be summarized, when compared with the conventional diffraction methods of both x-rays and neutrons.

(1) The structure factor can be determined in a very wide q region and thus the so-called termination error is almost eliminated in the Fourier transformation.

(2) Fluctuations in the source intensity do not affect the results, because of the parallel counting of photons with different energies.

(3) The diffraction angle is fixed during one measurement in the so-called variable energy mode. This results in free from danger of optical misalignment and thus a high reproducibility for in-situ measurement is possible.

(4) The integrated intensity of the continuous spectrum source is usually higher than the intensity of the conventional characteristic radiation or the neutrons selected by a monochromator crystal and thus a higher rate of data collection is easily obtained.

Contrary to these advantages, there are some disadvantages such as the Placzek correction for the departure from the static approximation in TOF pulsed neutron diffraction experiment and the limited resolution of the solid state detector in EDXD measurement. The experimental set-up must be chosen to minimize these factors. The fairly complex data processing mainly due to the energy dependence of various quantities in both techniques has been almost completed recently and thus the structural determination of disordered materials is now possible by these two techniques.

Figures (A2-2) and (A2-3) give the structure factor and the RDF obtained by these relatively new techniques using the results of metallic alloy glasses of $Pd_{80}Si_{20}$ (Suzuki et al. 1976) and $(Pt_{0.8}Ni_{0.2})_{73}P_{27}$ (Aur and Egami 1982) as an example. As is easily seen in these figures, the oscillations in the structure factor are clearly detected in the higher value of q. Figure (A2-2) is the RDF of the $Pd_{80}Si_{20}$ metallic glass obtained by the TOF pulsed neutron diffraction which also demonstrates the truncation effect on the Fourier

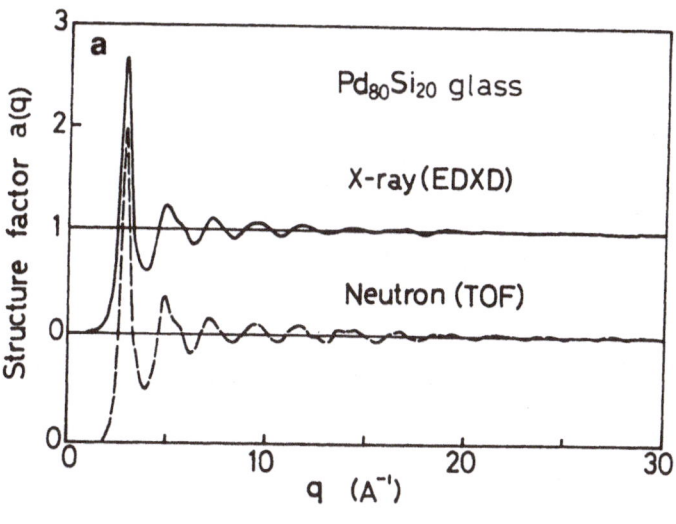

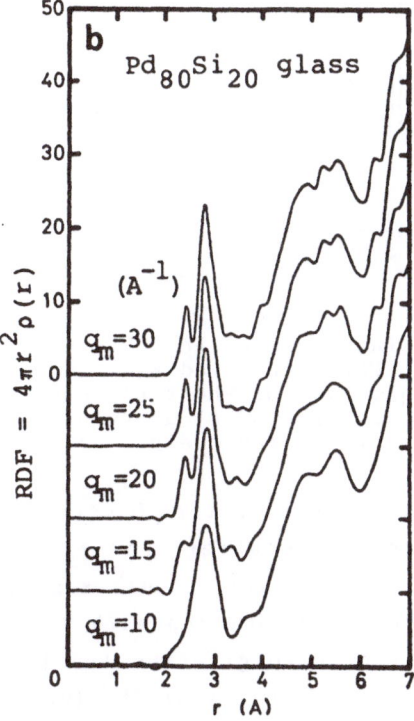

Fig.A2-2 Structure factor (a) and radial distribution function (b) of the $Pd_{80}Si_{20}$ glass obtained by TOF pulsed neutron diffraction (Suzuki et al. 1976). Structure factor obtained by EDXD (Tamaki and Waseda 1980).

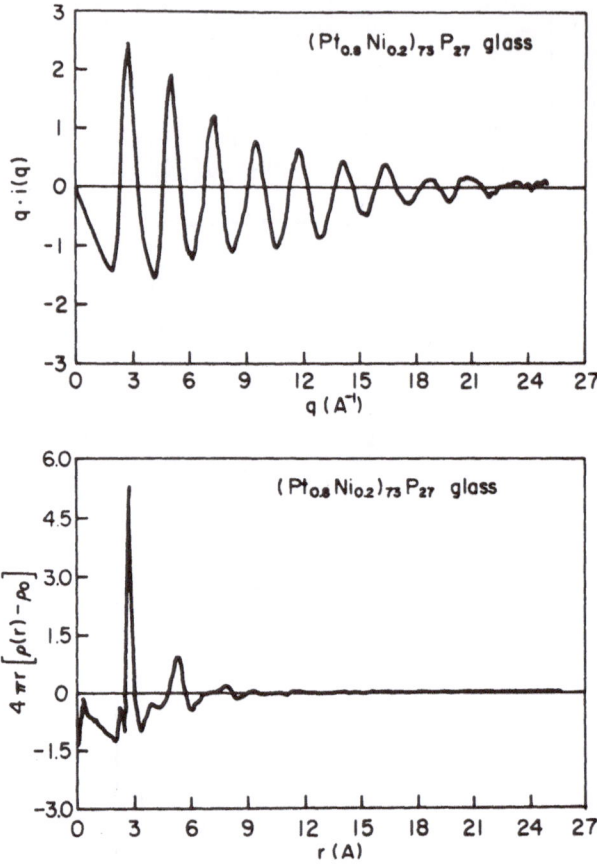

Fig.A2-3 Structural functions of the $(Pt_{0.8}Ni_{0.2})_{73}P_{27}$ glass
obtained by EDXD. (Aur and Egami 1980). The small
peak at inner side of the first peak corresponds to
the Pt-P pairs (r=2.32 A) un-resolved by conventional
x-ray diffraction previously.

transformation. A splitting in the first peak of RDF curve is
evidently observed when the Fourier transformation calculation is
truncated beyond q_{max} = 15 A^{-1}. This result suggests that the RDF data
obtained by the Fourier transformation with the relatively low q_{max}
value is not reliable enough to give the local ordering structure in
near neighbor region. In the results of the $(Pt_{0.8}Ni_{0.2})_{73}P_{27}$ metallic
glass determined by the EDXD method (see Fig.(A2-3)), the small peak
inside the first peak is detected. This small peak corresponds to the
correlation of Pt-P pairs which cannot be resolved by the previous
conventional x-ray diffraction.

The high wave vector measurement appears to be promising for

...igher resolution in RDF and thus permitting the possible separation of the pair correlations in near neighbor region from the **total structure factor alone,** i.e., without separation of partial structure factors. This is one way to provide valuable information on the local ordering structure of disordered materials in cases where no partial structural function is available. Of course, the information of partial functions are required for a detailed (quantitative) discussion on the atomic scale structure including higher ordering such as the second and third nearest neighbor correlations in multi-component disordered materials.

APPENDIX 3

Theory for the Scattering Intensity of a Particular System indicating both Crystal-like Laue-Bragg Peaks and Liquid-like Diffuse Patterns

The scattering theory on the structure of solid fast ion conductors exemplified by noble metal-chalcogenides such as Ag_2Se and Ag_2S has been discussed in detail (see for example, Hoshino 1957, Funke 1976, Tsuchiya et al. 1978). However, for convenience of future structural study of various conductors, we give here a brief background on the theoretical equations for x-ray scattering intensity of a particular system indicating both crystal-like Laue-Bragg peaks and liquid-like diffuse patterns.

The high ionic conductivity of solid fast ion conductors is attributed to the fast diffusion coefficient of cations (anions in a few cases such as PbF_2) which is known to be almost comparable to the self-diffusion coefficient in liquid metals (see for example, Okazaki 1967). Therefore, the following basic idea for the structure of solid fast ion conductors may be well recognized. The anions form a crystal-like periodic lattice, whereas the cations are supposed to distribute at random similar to liquids within the vacant space in each cell formed by the anions. The charge neutrality in each cell is maintained. By using common Fourier transformation as is applied for liquids and amorphous solids, we can evaluate the average atomic scale correlation functions of three pairs, anion-anion, cation-cation and anion-cation. They are, in principle, determined from measured intensity data without any structural model such as the size displacement effect (see for example, DeRidder et al. 1974).

Let us consider the coherent x-ray scattering intensity $I^{coh}(q)$ for a system containing N_A anions and N_B cations. It can be given by:

$$I^{coh}(\vec{q}) = \langle F(\vec{q}) \cdot F^*(\vec{q}) \rangle \tag{A3-1}$$

$$F(\vec{q}) = \sum_{k}^{N_A} f_A \exp(i\vec{q}\vec{r}_k) + \sum_{l}^{N_C} f_C \exp(i\vec{q}\vec{r}_l) \tag{A3-2}$$

where q is the so-called wave vector, f is the usual atomic scattering factor excluding its q-dependence for simplification and the asterisk denotes the complex conjugate. Putting eq.(A3-2) into eq.(A3-1), the following equation can be obtained.

$$I^{coh}(q) = I^{coh}_{AA}(q) + I^{coh}_{AB}(q) + I^{coh}_{BB}(q) \tag{A3-3}$$

where

$$I^{coh}_{AA}(q) = \langle f_A f_A^* \Sigma\Sigma \exp\{i\vec{q}\cdot(\vec{r}_k - \vec{r}_{k'})\}\rangle$$

$$I^{coh}_{AB}(q) = \langle f_A f_B^* \Sigma\Sigma \exp\{i\vec{q}\cdot(\vec{r}_k - \vec{r}_{l'})\}\rangle + \langle cc\rangle \tag{A3-4}$$

$$I^{coh}_{BB}(q) = \langle f_B f_B^* \Sigma\Sigma \exp\{i\vec{q}\cdot(\vec{r}_l - \vec{r}_{l'})\}\rangle$$

and $\langle cc\rangle$ corresponds the complex conjugate term.

The first term of $I^{coh}(q)$ in eq.(A3-3) yields only crystal-like Laue-Bragg peaks written as follows:

$$N_A f_A f_A^* + N_A(N_A-1)f_A f_A^* \delta_{\vec{q},\vec{G}} = N_A f_A f_A^* + N^2 x^2 f_A f_A^* \delta_{\vec{q},\vec{G}} \tag{A3-5}$$

where $N = N_A + N_B$, δ is the so-called Kronecker symbol, G is the reciprocal lattice vector specified by the anion structure and x represents the concentration of anions given by N_A/N. The second term of $I^{coh}(q)$ in eq.(A3-3) may be rewritten as follows.

$$f_A f_B^* \langle \overset{N_A}{\Sigma}\exp(i\vec{q}\cdot\vec{r})\rangle\langle\overset{N_B}{\Sigma}\exp(-i\vec{q}\cdot\vec{r})\rangle + f_A f_B^* \langle\overset{N_A N_B}{\Sigma\Sigma}\exp\{i\vec{q}\cdot(\vec{r}_k - \vec{r}_{l'})\}\rangle$$

$$- \langle\overset{N_A}{\Sigma}\exp(i\vec{q}\cdot\vec{r}_k)\rangle\langle\overset{N_B}{\Sigma}\exp(-i\vec{q}\cdot\vec{r}_l)\rangle + \langle cc\rangle \tag{A3-6}$$

The anions form a crystal-like lattice and then it follows that :

$$\langle\overset{N_B}{\Sigma}\exp(-i\vec{q}\cdot\vec{r}_l)\rangle = \frac{N_B}{N_A}\overset{N_A}{\Sigma}\exp(-i\vec{q}\cdot\vec{r}_k)\langle\exp(-i\vec{q}\cdot\vec{\rho})\rangle \tag{A3-7}$$

Since the configurational average of $\exp(-i\vec{q}\cdot\vec{\rho}_{k'l'})$ is independent of $k'l'$, the suffix for $\vec{\rho}$ in eq.(A3-7) can be omitted. Substituting eq. (A3-7) in the first term of eq.(A3-6), we obtain:

$$f_A f_B^* N_A N_B \delta_{\vec{q},\vec{G}}\langle\exp(-i\vec{q}\cdot\vec{\rho})\rangle + \langle cc\rangle$$

$$= N^2 x(1-x)\left[f_A f_B^* \langle\exp(-i\vec{q}\cdot\vec{r})\rangle + f_A^* f_B \langle\exp(i\vec{q}\cdot\vec{r})\rangle\right]\delta_{\vec{q},\vec{G}} \tag{A3-8}$$

This term can be detected as usual Laue-Bragg peaks modulated by the factor of $\langle \exp(-i\vec{q}\cdot\vec{r})\rangle$.

By introducing the partial pair distribution function between anion and cation, $g_{AB}(\vec{r}_k-\vec{r}_l)$, the last two brackets of eq.(A3-7) can be expressed as follows.

$$f_A f_B^* \sum_{A}^{N_A} \sum_{B}^{N_B} \exp\{i\vec{q}\cdot(\vec{r}-\vec{r})\}\left[g_{AB}(\vec{r}_k-\vec{r}_l)-1\right]d\vec{r} \ /V + (cc)$$

$$= x(1-x)N(f_A f_B^* + f_A^* f_B)\left[\tilde{S}_{AB}(\vec{q})-1\right] \qquad (A3-9)$$

where V is the volume of a system and $\tilde{S}_{AB}(\vec{q})$ indicates the partial structure factor of anion and cation pairs which corresponds to the Fourier transform of $g_{AB}(\vec{r}_k-\vec{r}_l)$. Since anions form a periodic lattice, the average distribution of cations may have the same periodicity as the anion lattice and yields the Laue-Bragg peaks as given by eq.(A3-7). However, it should be remembered that the cation distribution is considered to be continuous, in a sense, except for the positions of anions, and thus the distribution function of $g_{AB}(\vec{r}_k-\vec{r}_l)$ is not necessarily periodic.

The third term of $I_{BB}^{coh}(q)$ in eq.(A3-3) can be also rewritten as follows in terms of the structure factor of cation-cation pairs, $\tilde{S}_{BB}(\vec{q})$.

$$f_B f_B^* \langle \sum_{k}^{N_B} \exp(i\vec{q}\cdot\vec{r}_k)\rangle\langle\sum_{k'}^{N_B}\exp(-i\vec{q}\cdot\vec{r}_{k'})\rangle$$

$$+ f_B f_B^* \langle \sum_{k}^{N_B}\sum_{k}^{N_B} \exp\{i\vec{q}\cdot(\vec{r}_k-\vec{r}_{k'})\rangle - \langle\sum^{N_B}\exp(-i\vec{q}\cdot\vec{r}_k)\rangle\langle\sum^{N_B}\exp(-i\vec{q}\cdot\vec{r}_{k'})\rangle$$

$$= (1-x)Nf_B f_B^* + N^2(1-x)^2 f_B f_B^* \delta_{\vec{q},\vec{G}} \langle \exp(i\vec{q}\cdot\rho)\rangle\langle\exp(-i\vec{q}\cdot\rho)\rangle$$

$$+ f_B f_B^* N(1-x)^2\left[\tilde{S}_{BB}(\vec{q})-1\right] \qquad (A3-10)$$

where $\tilde{S}_{BB}(\vec{q})$ is given by the Fourier transform of the partial pair distribution function between cation pairs.

The partial structure factors $\tilde{S}_{AB}(\vec{q})$ and $\tilde{S}_{BB}(\vec{q})$ for solid fast ion conductors generally depend on the direction of scattering vector q. Such anisotropic structural information for anion-cation and cation-cation pairs could be obtained directly from the measurement of a single crystal sample. In the case of a polycrystalline or powder sample, the angular dependence of $\tilde{S}_{AB}(\vec{q})$ or $\tilde{S}_{BB}(\vec{q})$ is averaged out and the isotropic partial structure factors as observed in liquids may be well recognized, and then average partial pair distribution function $\tilde{g}_{ij}(r)$ can evaluated by the following well-known equation.

$$\tilde{g}_{ij}(r) = 1 + \frac{V}{2\pi^2(N_A+N_B)} \int_0^\infty q^2 \left[\tilde{S}_{ij}(q)-1\right] \frac{\sin(q \cdot r)}{q \cdot r} \, dq \qquad (A3-11)$$

A number of diffraction studies for solid fast ion conductors indicate that the motion of anions may be approximated as independent Einstein vibration (see for example, Hoshino et al. 1977). From this point of view, the equations for $I^{coh}(q)$ should involve the thermal vibration effect expressed by the so-called Debye-Waller type factor such as $f_A \exp(-B_A\vec{q}^2/16\pi^2)$. Therefore, the observed coherent x-ray scattering intensity for a polycrystalline sample of solid fast ion conductor can be expressed, in the first approximation, by the following equation.

$$\begin{aligned}
I^{coh}_{obs}(q) &= N^2\{xf_A\exp(-B_A \cdot \vec{q}^2/16\pi^2) + (1-x)<\exp(i\vec{q}\cdot\rho)>f_B\} \\
&\quad \times \{xf_A^*\exp(-B_A \cdot \vec{q}^2/16\pi^2) + (1-x)<\exp(i\vec{q}\cdot\rho)>f_B^*\}\delta_{\vec{q},\vec{G}} \\
&\quad + Nxf_Af_A^*\{1-\exp(-2B_A\vec{q}^2/16\pi^2) + (1-x)f_Bf_B^*\} \\
&\quad + (f_Af_B^*+f_A^*f_B)x(1-x)\left[\tilde{S}_{AB}(q)-1\right] + (1-x)^2f_Bf_B^*\left[\tilde{S}_{BB}(q)-1\right] \quad (A3-12)
\end{aligned}$$

The first term of eq.(A3-12) contributes to the normal Laue-Bragg peaks and the second term to the liquid-like diffuse patterns. It is worth mentioning that the latter contains the contribution from both coherent scattering between anion-cation and cation-cation pairs. Equation (A3-12) also indicates that the observed diffuse component can be decomposed into the two partial structure factors of anion-cation and cation-cation pairs by the standard techniques frequently used for the structural study of disordered materials. The anomalous x-ray scattering technique is, of course, one way to separate the partial structure factors $\tilde{S}_{AB}(\vec{q})$ and $\tilde{S}_{BB}(\vec{q})$ from measured diffuse patterns, and thus it is possible to evaluate the average information on the atomic scale structure as a function of the relative distance between anion and cation or cation and cation pairs through eq.(A3-11).

Based on some experimental results (see for example, Boyce et al. 1977), it is more realistic to include the vibrational effect of cation, and also anion to the partial structural functions and the phase factor of $<\exp(-i\vec{q}\cdot\vec{r})>$ in the diffuse scattering terms in $I^{coh}(q)$. The anharmonic thermal vibration of the constituents may be also taken into account for some cases. The evaluation of these components is considered to be significant in model calculation on the structure of solid fast ion conductors. However, such effects might be automatically included in the information of $\tilde{S}_{AB}(\vec{q})$, $\tilde{S}_{BB}(\vec{q})$ and $<\exp(-i\vec{q}\cdot\vec{r})>$ determined from measured x-ray scattering intensity. For neutron diffraction experiment for solid fast ion conductors, the atomic scattering factor f_n employed in this appendix should be replaced by b_n, the nuclear scattering length (Bacon 1969, 1972).

APPENDIX 4

Energy Dependence (1-50 keV) of X-ray Anomalous Dispersion Factors f' and f" calculated by the Cromer-Liberman's Scheme

X-ray anomalous dispersion factors f' and f" for a total of 96 neutral atoms have recently been evaluated in the energy range between 1 and 50 keV (Waseda et al. 1984). The basic procedures of calculation in this work followed a method proposed by Cromer and Liberman (1970); i.e., the relativistic estimation of f' and f" was made numerically without approximation to the form of the cross section vs energy curve given by Cromer and Liberman (Los Alamos Scientific Laboratory Report, LA4403 in 1970). The present calculations were made with 48 points in the Gaussian integral at intervals of 1 eV in the close vicinity of the absorption edge. However, the following point may be suggested. The calculations at smaller intervals such as 0.1 eV indicate more distinct change in the real part of f'. Numerical examples are -30.875 for E=5.0116 keV, in comparison with -22.519 for E=5.011 keV and -24.266 for E=5.012 keV in the case of cesium. Absorption edges of K, L and M series for neutral atoms in keV unit obtained in this calculation are also summarized in **Table A4-1**. For convenience of future investigation by the anomalous x-ray scattering (AXS) and the energy dispersive x-ray diffraction (EDXD) techniques, this appendix provides the energy dependence of x-ray anomalous dispersion factors in the graphic form. More detailed numerical information including mass absorption coefficients as incidental data may be available upon request.

This calculation was supported in part by the Ito Science Foundation and the Toshiba Corporate Research and Development Center (Kawasaki) through grants to the author.

Table A4-1 Absorption edges of K, L and M series for neutral atoms in keV unit obtained in the present calculation.

	K	L_{III}	L_{II}	L_I	M_V	M_{IV}	M_{III}	M_{II}	M_I	
11 Na	1.073									11 Na
12 Mg	1.305									12 Mg
13 Al	1.560									13 Al
14 Si	1.839									14 Si
15 P	2.146									15 P
16 S	2.472									16 S
17 Cl	1.823									17 Cl
18 Ar	3.203									18 Ar
19 K	3.608									19 K
20 Ca	4.039									20 Ca
21 Sc	4.493									21 Sc
22 Ti	4.967									22 Ti
23 V	5.466									23 V
24 Cr	5.990									24 Cr
25 Mn	6.539									25 Mn
26 Fe	7.112									26 Fe
27 Co	7.709									27 Co
28 Ni	8.333			1.009						28 Ni
29 Cu	8.979			1.097						29 Cu
30 Zn	9.659	1.020	1.043	1.194						30 Zn
31 Ga	10.367	1.116	1.143	1.298						31 Ga
32 Ge	11.103	1.217	1.248	1.415						32 Ge
33 As	11.867	1.324	1.359	1.527						33 As
34 Se	12.658	1.436	1.477	1.654						34 Se
35 Br	13.474	1.550	1.596	1.782						35 Br
36 Kr	14.326	1.675	1.728	1.921						36 Kr
37 Rb	15.200	1.805	1.864	2.066						37 Rb
38 Sr	16.105	1.940	2.007	2.217						38 Sr
39 Y	17.038	2.080	2.156	2.373						39 Y
40 Zr	17.998	2.223	2.307	2.532						40 Zr
41 Nb	18.986	2.371	2.465	2.698						41 Nb
42 Mo	20.000	2.521	2.626	2.866						42 Mo
43 Tc	21.044	2.677	2.794	3.043						43 Tc
44 Ru	22.117	2.838	2.967	3.224						44 Ru
45 Rh	23.220	3.004	3.147	3.412						45 Rh
46 Pd	24.350	3.174	3.331	3.605						46 Pd
47 Ag	25.514	3.352	3.524	3.806						47 Ag
48 Cd	26.711	3.538	3.727	4.018						48 Cd
49 In	27.940	3.731	3.938	4.238						49 In
50 Sn	29.200	3.929	4.157	4.465						50 Sn
51 Sb	30.491	4.133	4.381	4.699						51 Sb
52 Te	31.814	4.342	4.612	4.940					1.006	52 Te
53 I	33.169	4.558	4.853	5.189					1.073	53 I
54 Xe	34.561	4.783	5.104	5.453					1.145	54 Xe
55 Cs	35.982	5.012	5.360	5.714				1.065	1.217	55 Cs

Table A4-1 (continued)

	K	L_{III}	L_{II}	L_I	M_V	M_{IV}	M_{III}	M_{II}	M_I	
56 Ba	37.438	5.247	5.624	5.989			1.063	1.137	1.293	56 Ba
57 La	38.922	5.483	5.891	6.262			1.124	1.205	1.362	57 La
58 Ce	40.441	5.724	6.164	6.549			1.186	1.273	1.435	58 Ce
59 Pr	41.988	5.964	6.440	6.835			1.243	1.338	1.511	59 Pr
60 Nd	43.566	6.208	6.722	7.120			1.298	1.403	1.576	60 Nd
61 Pm	45.189	6.459	7.013	7.428	1.027	1.052	1.357	1.472	1.647	61 Pm
62 Sm	46.831	6.716	7.312	7.737	1.090	1.106	1.420	1.541	1.723	62 Sm
63 Eu	48.516	6.977	7.617	8.052	1.131	1.161	1.481	1.614	1.800	63 Eu
64 Gd	50.236	7.243	7.930	8.376	1.186	1.218	1.544	1.688	1.881	64 Gd
65 Tb	51.993	7.514	8.252	8.708	1.242	1.275	1.612	1.768	1.968	65 Tb
66 Dy	53.785	7.790	8.581	9.046	1.295	1.333	1.676	1.842	2.047	66 Dy
67 Ho	55.615	8.071	8.918	9.394	1.352	1.392	1.742	1.923	2.129	67 Ho
68 Er	57.482	8.358	9.264	9.751	1.410	1.454	1.812	2.006	2.207	68 Er
69 Tm	59.386	8.648	9.617	10.115	1.468	1.515	1.885	2.090	2.307	69 Tm
70 Yb	61.329	8.944	9.978	10.486	1.528	1.577	1.950	2.173	2.398	70 Yb
71 Lu	63.310	9.244	10.348	10.870	1.589	1.640	2.024	2.264	2.492	71 Lu
72 Hf	65.347	9.561	10.739	11.270	1.662	1.717	2.108	2.366	2.601	72 Hf
73 Ta	67.412	9.881	11.135	11.681	1.735	1.794	2.194	2.469	2.708	73 Ta
74 W	69.521	10.206	11.543	12.099	1.810	1.872	2.281	2.575	2.820	74 W
75 Re	71.672	10.535	11.958	12.526	1.883	1.949	2.368	2.682	2.932	75 Re
76 Os	73.866	10.870	12.384	12.967	1.960	2.031	2.458	2.792	3.049	76 Os
77 Ir	76.107	11.215	12.823	13.418	2.041	2.116	2.551	2.909	3.174	77 Ir
78 Pt	78.390	11.563	13.272	13.879	2.122	2.202	2.646	3.027	3.296	78 Pt
79 Au	80.720	11.918	13.733	14.351	2.206	2.291	2.743	3.148	3.425	79 Au
80 Hg	83.097	12.283	14.208	14.838	2.295	2.385	2.847	3.279	2.562	80 Au
81 Tl	85.525	12.657	14.697	15.346	2.390	2.485	2.957	3.416	3.704	81 Tl
82 Pb	87.999	13.034	15.119	15.860	2.484	2.586	3.067	3.554	3.851	82 Pb
83 Bi	90.521	13.418	15.710	16.387	2.580	2.688	3.177	3.697	3.999	83 Bi
84 Po	93.100	13.813	16.243	16.938	2.683	2.798	3.302	3.854	4.150	84 Po
85 At	95.724	14.213	16.784	17.492	2.787	2.909	3.426	4.008	4.317	85 At
86 Rn	98.398	14.619	17.336	18.048	2.893	3.022	3.538	4.159	4.482	86 Rn
87 Fr	101.130	15.030	17.905	18.638	3.000	3.136	3.663	4.327	4.652	87 Fr
88 Ra	103.920	15.443	18.483	19.236	3.105	3.249	3.792	4.490	4.822	88 Ra
89 Ac	106.750	15.870	19.082	19.839	3.219	3.370	3.909	4.656	5.002	89 Ac
90 Th	109.640	16.299	19.692	20.471	3.332	3.491	4.046	4.832	5.182	90 Th
91 Pa	112.590	16.732	20.313	21.103	3.442	3.611	4.174	5.001	5.367	91 Pa
92 U	115.600	17.165	20.946	21.756	3.552	3.728	4.304	5.182	5.548	92 U
93 Np	118.670	17.609	21.599	22.425	3.666	3.851	4.435	5.366	5.723	93 Np
94 Pu	121.810	18.056	22.265	23.096	3.708	3.973	4.557	5.541	5.933	94 Pu
95 Am	125.020	18.503	22.943	23.772	3.887	4.092	4.667	5.710	6.121	95 Am
96 Cm	128.210	18.929	23.778	24.459	3.971	4.227	4.797	5.895	6.288	96 Cm
97 Bk	131.580	19.451	24.384	25.274	4.132	4.366	4.977	6.147	6.556	97 Bk
98 Cf	135.950	19.929	25.249	26.108	4.253	4.497	5.109	6.359	6.754	98 Cf

The experimental information of the absorption edges for various elements is given in detail by J.A.Bearden:(Rev. Mod. Phys. 39(1967),78).

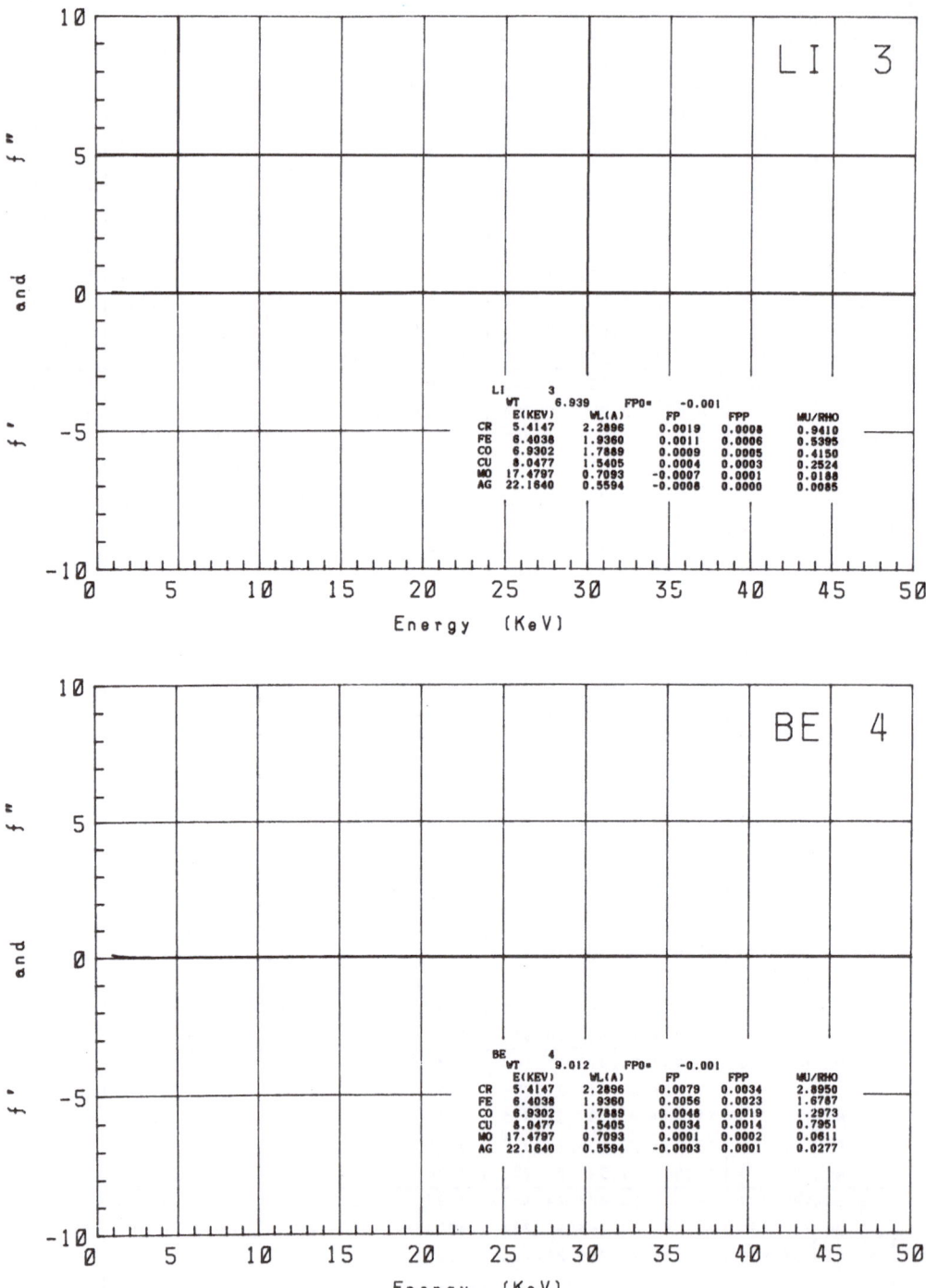

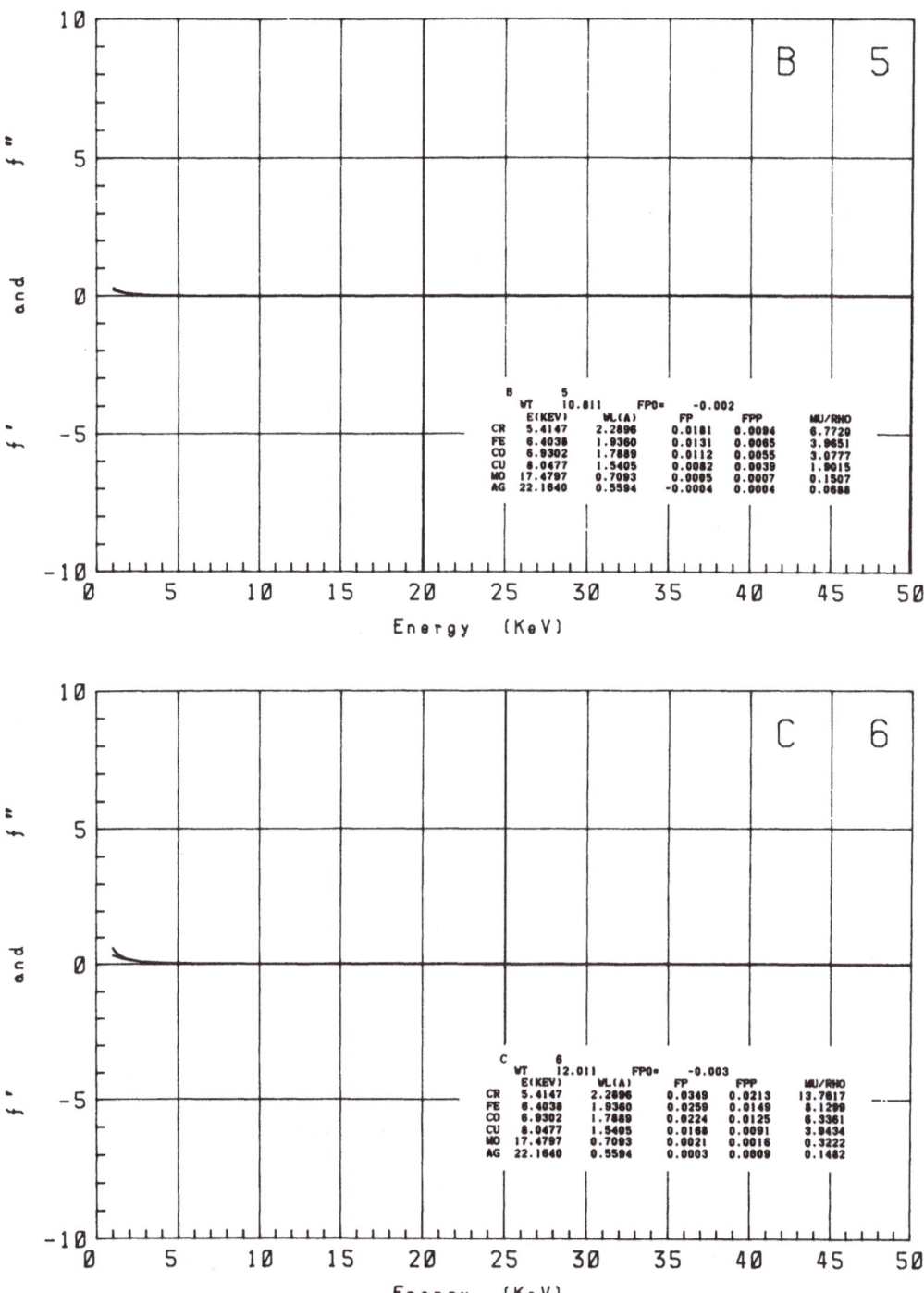

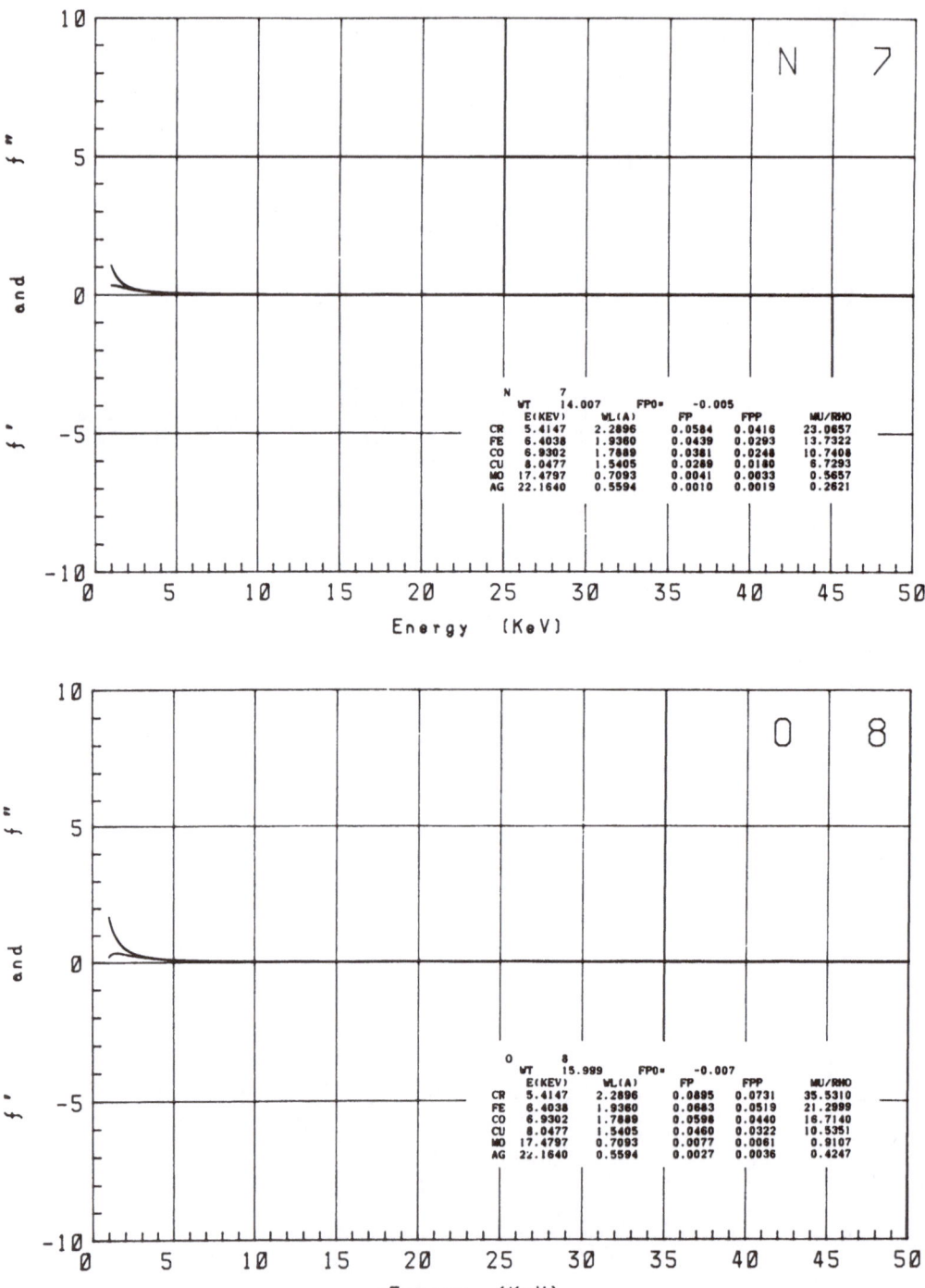

N   7
WT   14.007   FPO=   -0.005

	E(KEV)	WL(A)	FP	FPP	MU/RHO
CR	5.4147	2.2896	0.0584	0.0416	23.0657
FE	6.4038	1.9360	0.0439	0.0293	13.7322
CO	6.9302	1.7889	0.0381	0.0248	10.7408
CU	8.0477	1.5405	0.0289	0.0180	6.7293
MO	17.4797	0.7093	0.0041	0.0033	0.5657
AG	22.1640	0.5594	0.0010	0.0019	0.2621

O   8
WT   15.999   FPO=   -0.007

	E(KEV)	WL(A)	FP	FPP	MU/RHO
CR	5.4147	2.2896	0.0895	0.0731	35.5310
FE	6.4038	1.9360	0.0683	0.0519	21.2999
CO	6.9302	1.7889	0.0598	0.0440	16.7140
CU	8.0477	1.5405	0.0460	0.0322	10.5351
MO	17.4797	0.7093	0.0077	0.0061	0.9107
AG	22.1640	0.5594	0.0027	0.0036	0.4247

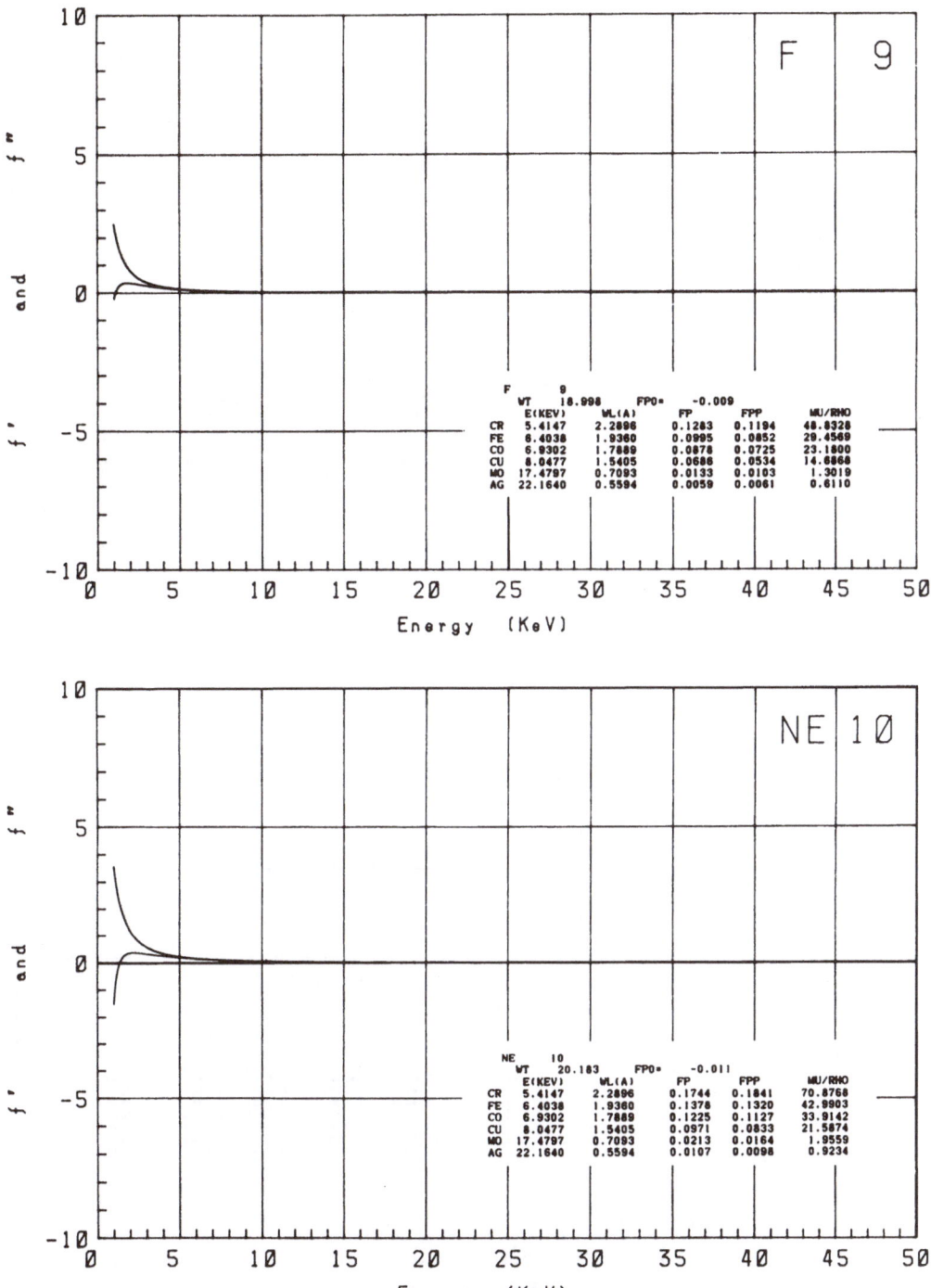

F 9

F	9			
WT 18.998	FPO=	-0.009		
E(KEV)	WL(A)	FP	FPP	MU/RHO
CR	5.4147	2.2896	0.1283	0.1194	48.8328
FE	6.4038	1.9360	0.0995	0.0852	29.4569
CO	6.9302	1.7889	0.0878	0.0725	23.1800
CU	8.0477	1.5405	0.0686	0.0534	14.6868
MO	17.4797	0.7093	0.0133	0.0103	1.3019
AG	22.1640	0.5594	0.0059	0.0061	0.6110

Energy (KeV)

NE 10

NE	10			
WT 20.183	FPO=	-0.011		
E(KEV)	WL(A)	FP	FPP	MU/RHO
CR	5.4147	2.2896	0.1744	0.1641	70.8768
FE	6.4038	1.9360	0.1378	0.1320	42.9903
CO	6.9302	1.7889	0.1225	0.1127	33.9142
CU	8.0477	1.5405	0.0971	0.0833	21.5874
MO	17.4797	0.7093	0.0213	0.0164	1.9559
AG	22.1640	0.5594	0.0107	0.0098	0.9234

Energy (KeV)

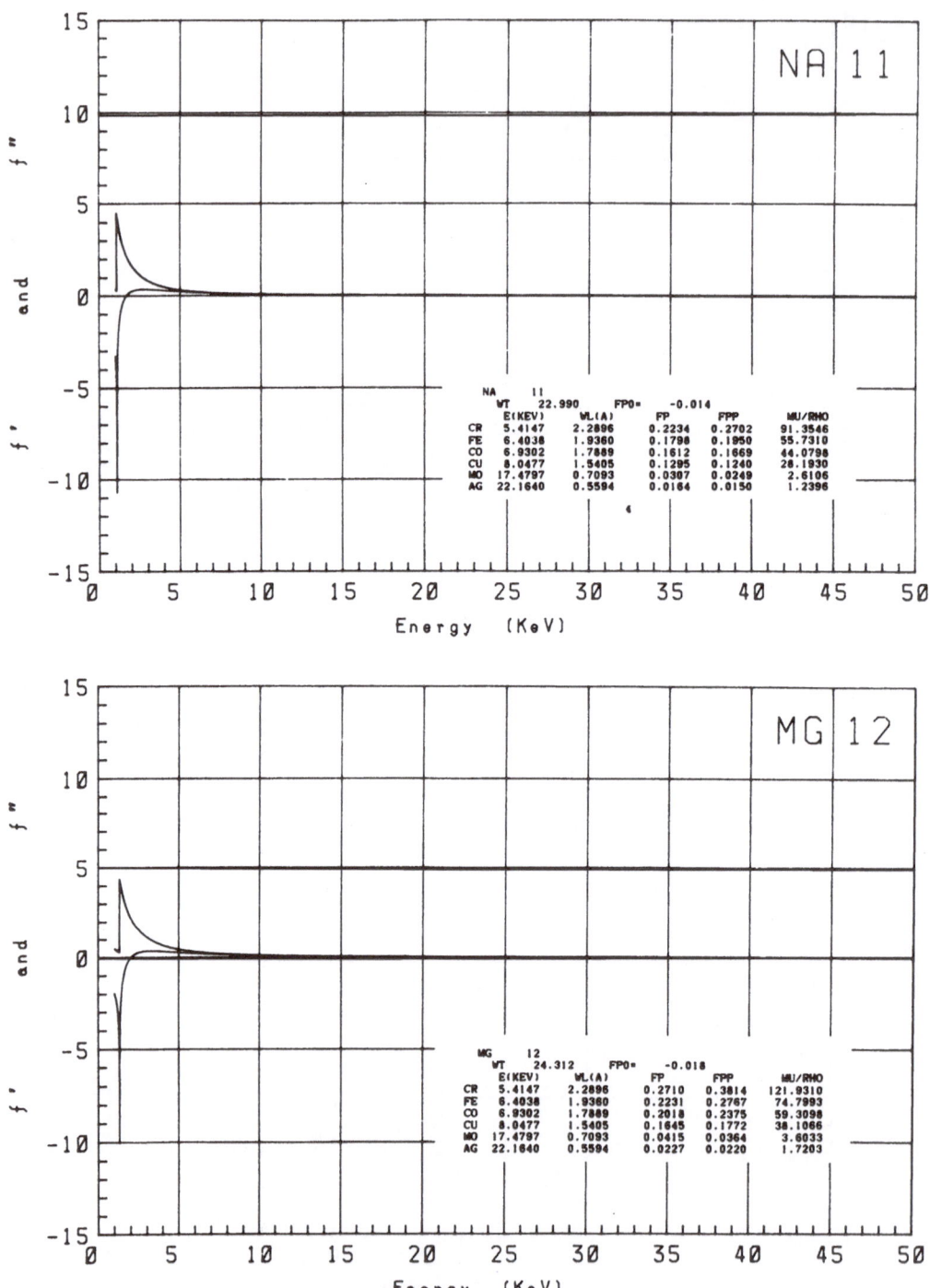

NA 11

NA 11
WT 22.990 FPO= -0.014

	E(KEV)	WL(A)	FP	FPP	MU/RHO
CR	5.4147	2.2896	0.2234	0.2702	91.3546
FE	6.4038	1.9360	0.1798	0.1950	55.7310
CO	6.9302	1.7889	0.1612	0.1669	44.0798
CU	8.0477	1.5405	0.1295	0.1240	28.1930
MO	17.4797	0.7093	0.0307	0.0249	2.6106
AG	22.1640	0.5594	0.0164	0.0150	1.2396

MG 12

MG 12
WT 24.312 FPO= -0.018

	E(KEV)	WL(A)	FP	FPP	MU/RHO
CR	5.4147	2.2896	0.2710	0.3814	121.9310
FE	6.4038	1.9360	0.2231	0.2767	74.7993
CO	6.9302	1.7889	0.2018	0.2375	59.3098
CU	8.0477	1.5405	0.1645	0.1772	38.1066
MO	17.4797	0.7093	0.0415	0.0364	3.6033
AG	22.1640	0.5594	0.0227	0.0220	1.7203

Energy (KeV)

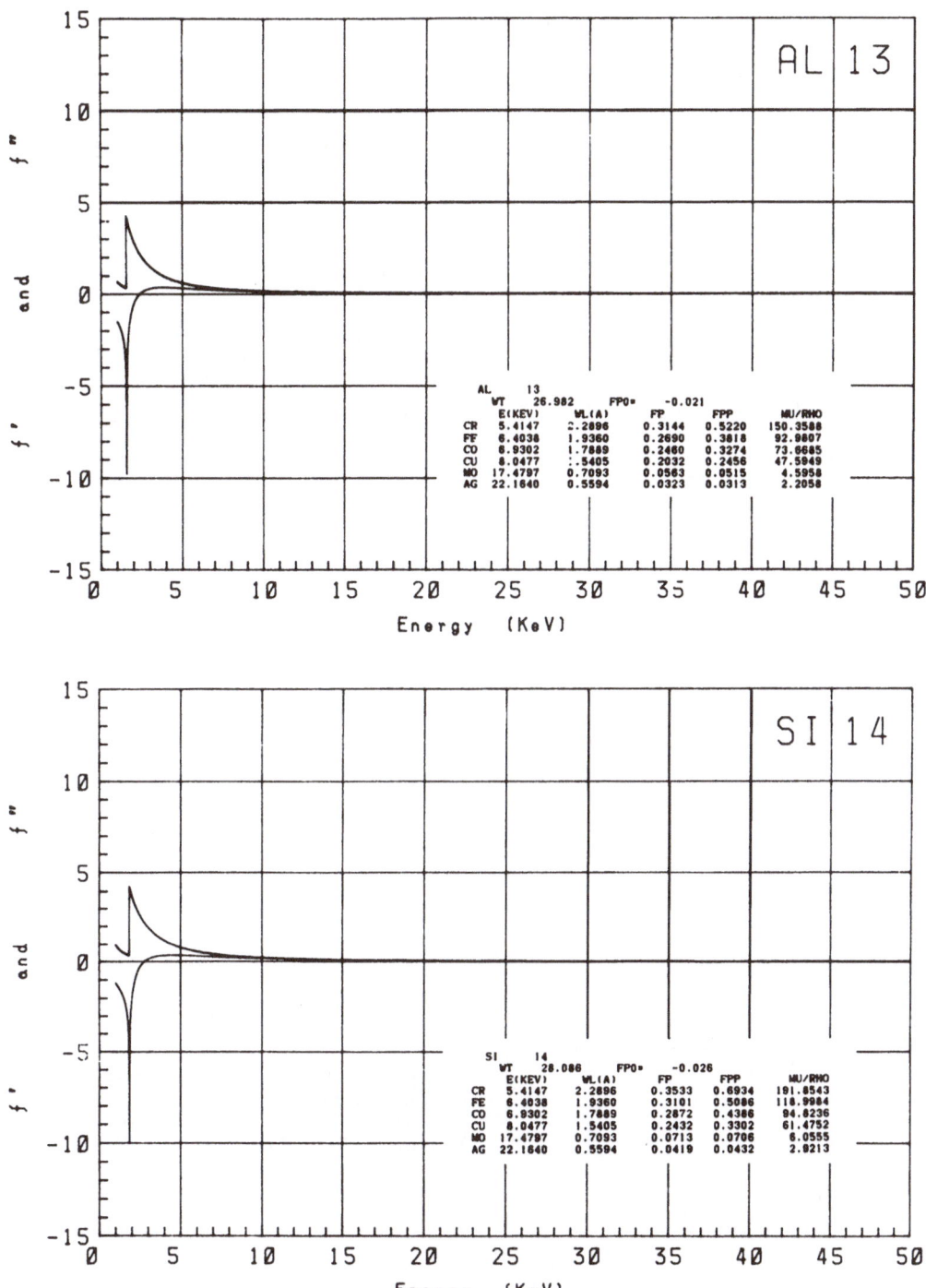

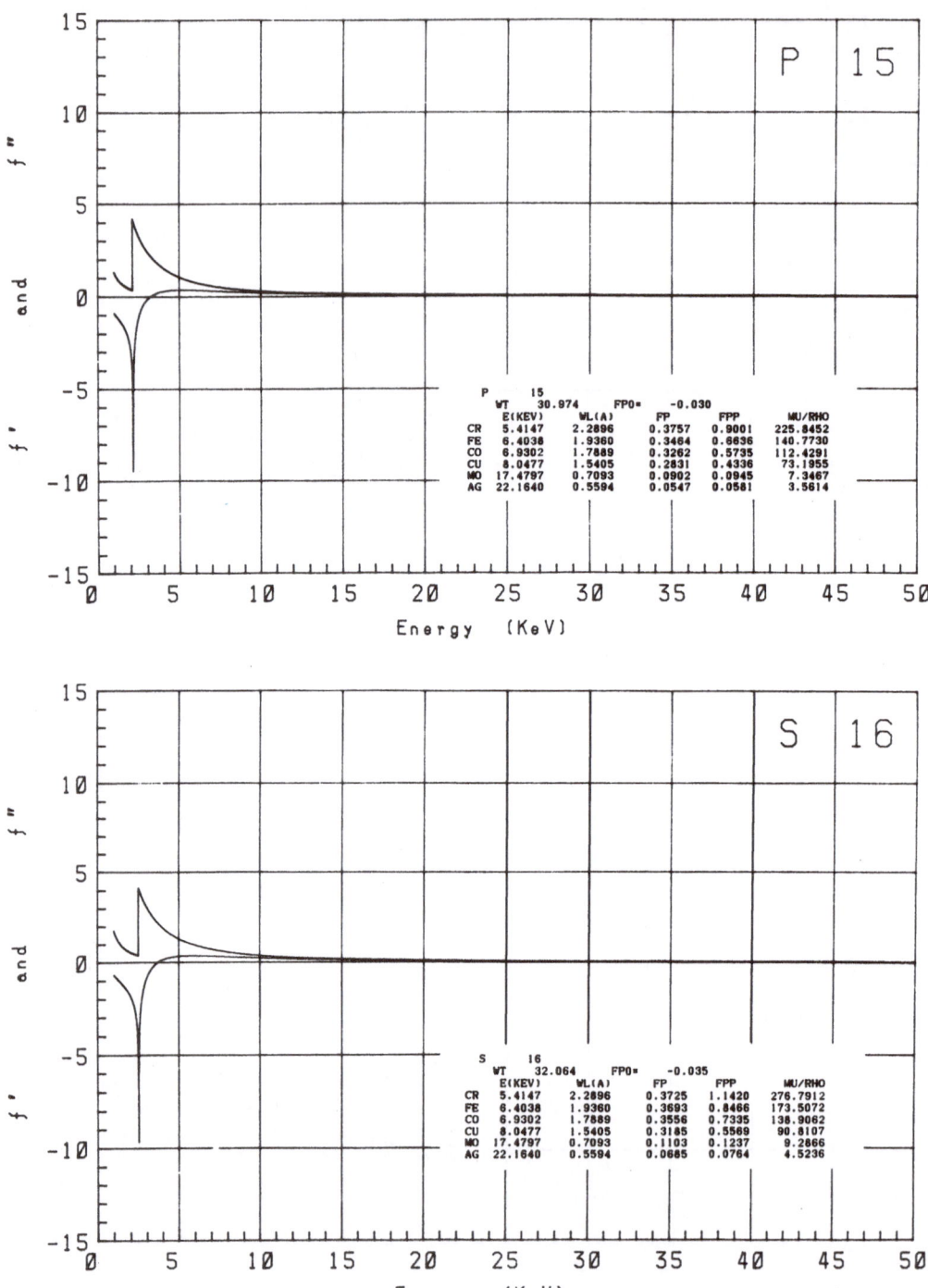

P 15

	E(KEV)	WL(A)	FP	FPP	MU/RHO
	WT	30.974	FPO=	-0.030	
CR	5.4147	2.2896	0.3757	0.9001	225.8452
FE	6.4038	1.9360	0.3464	0.6636	140.7730
CO	6.9302	1.7889	0.3262	0.5735	112.4291
CU	8.0477	1.5405	0.2831	0.4336	73.1955
MO	17.4797	0.7093	0.0902	0.0945	7.3467
AG	22.1640	0.5594	0.0547	0.0581	3.5614

Energy (KeV)

S 16

	E(KEV)	WL(A)	FP	FPP	MU/RHO
	WT	32.064	FPO=	-0.035	
CR	5.4147	2.2896	0.3725	1.1420	276.7912
FE	6.4038	1.9360	0.3693	0.8466	173.5072
CO	6.9302	1.7889	0.3556	0.7335	138.9062
CU	8.0477	1.5405	0.3185	0.5569	90.8107
MO	17.4797	0.7093	0.1103	0.1237	9.2866
AG	22.1640	0.5594	0.0685	0.0764	4.5236

Energy (KeV)

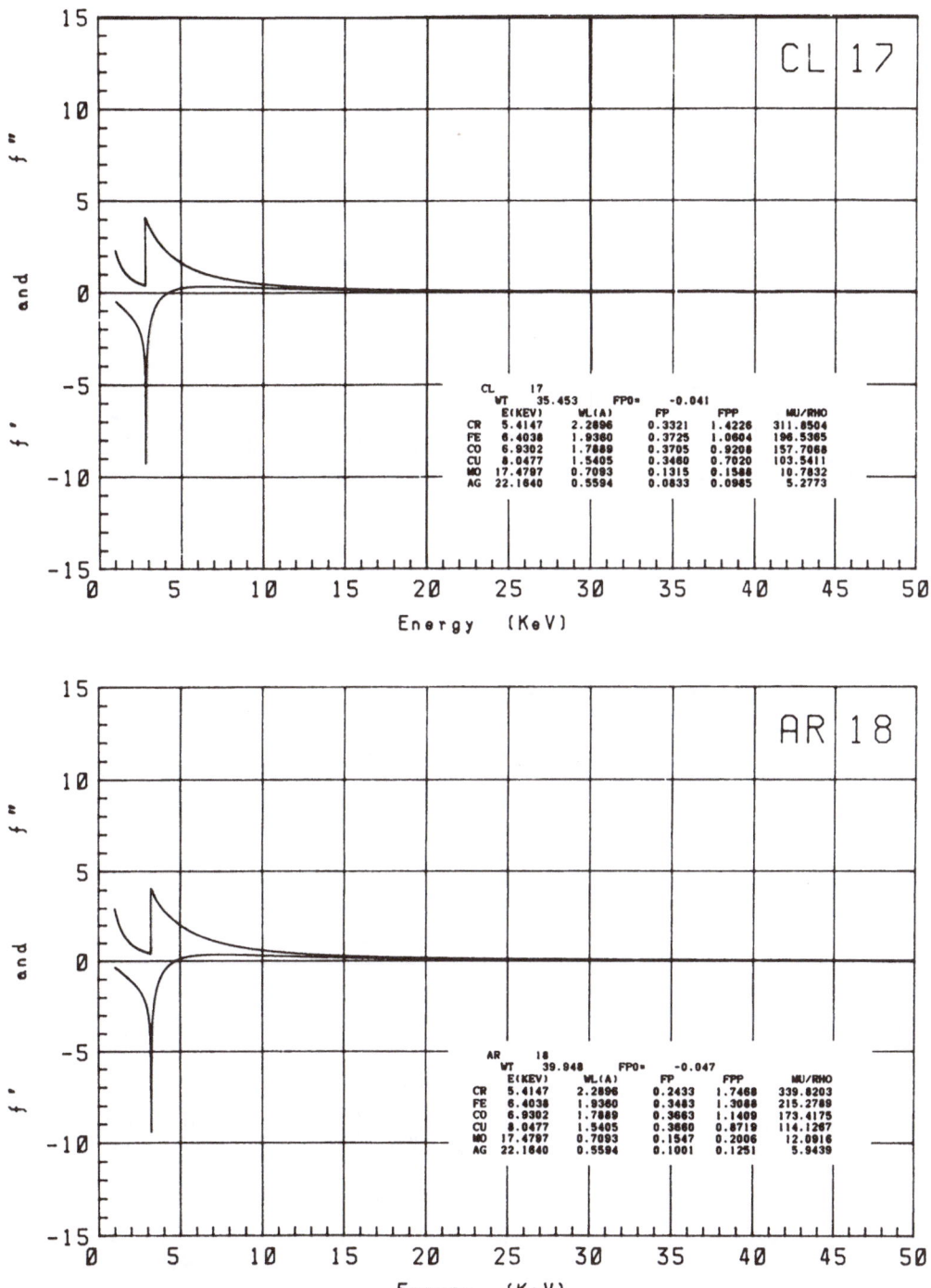

CL 17

CL   17
WT   35.453   FPO=   -0.041
	E(KEV)	WL(A)	FP	FPP	MU/RHO
CR	5.4147	2.2896	0.3321	1.4226	311.8504
FE	6.4038	1.9360	0.3725	1.0604	196.5365
CO	6.9302	1.7889	0.3705	0.9208	157.7068
CU	8.0477	1.5405	0.3460	0.7020	103.5411
MO	17.4797	0.7093	0.1315	0.1588	10.7832
AG	22.1640	0.5594	0.0833	0.0985	5.2773

AR 18

AR   18
WT   39.948   FPO=   -0.047
	E(KEV)	WL(A)	FP	FPP	MU/RHO
CR	5.4147	2.2896	0.2433	1.7468	339.8203
FE	6.4038	1.9360	0.3483	1.3088	215.2789
CO	6.9302	1.7889	0.3663	1.1409	173.4175
CU	8.0477	1.5405	0.3660	0.8719	114.1267
MO	17.4797	0.7093	0.1547	0.2006	12.0916
AG	22.1640	0.5594	0.1001	0.1251	5.9439

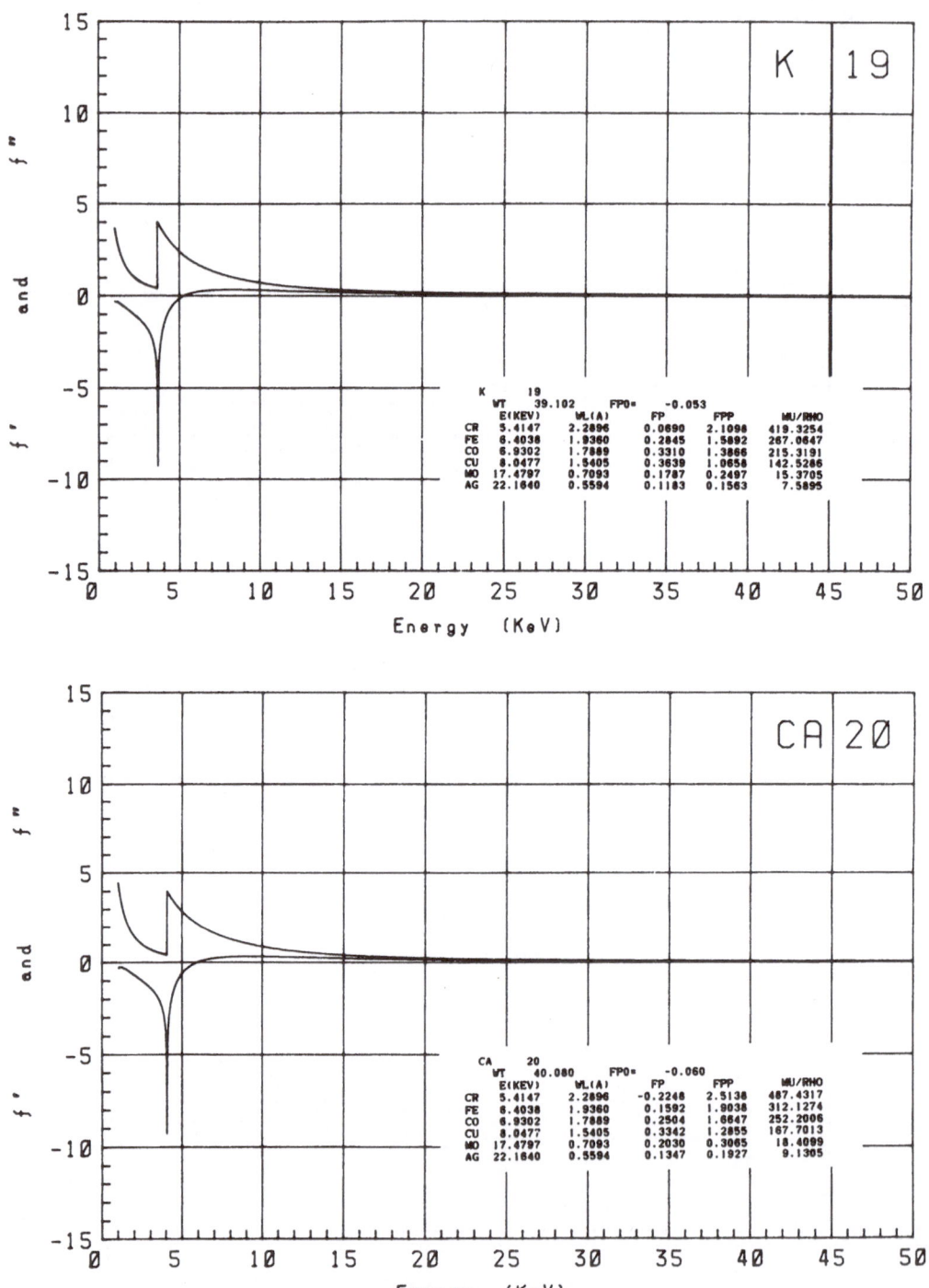

K 19

K	19			
WT	39.102	FPO=	-0.053	
	E(KEV)	WL(A)	FP	FPP	MU/RHO
CR	5.4147	2.2896	0.0690	2.1098	419.3254
FE	6.4038	1.9360	0.2845	1.5892	267.0647
CO	6.9302	1.7889	0.3310	1.3866	215.3191
CU	8.0477	1.5405	0.3639	1.0658	142.5286
MO	17.4797	0.7093	0.1787	0.2497	15.3705
AG	22.1640	0.5594	0.1183	0.1563	7.5895

CA 20

CA	20			
WT	40.080	FPO=	-0.060	
	E(KEV)	WL(A)	FP	FPP	MU/RHO
CR	5.4147	2.2896	-0.2246	2.5138	487.4317
FE	6.4038	1.9360	0.1592	1.9038	312.1274
CO	6.9302	1.7889	0.2504	1.6647	252.2006
CU	8.0477	1.5405	0.3342	1.2855	167.7013
MO	17.4797	0.7093	0.2030	0.3085	18.4099
AG	22.1640	0.5594	0.1347	0.1927	9.1305

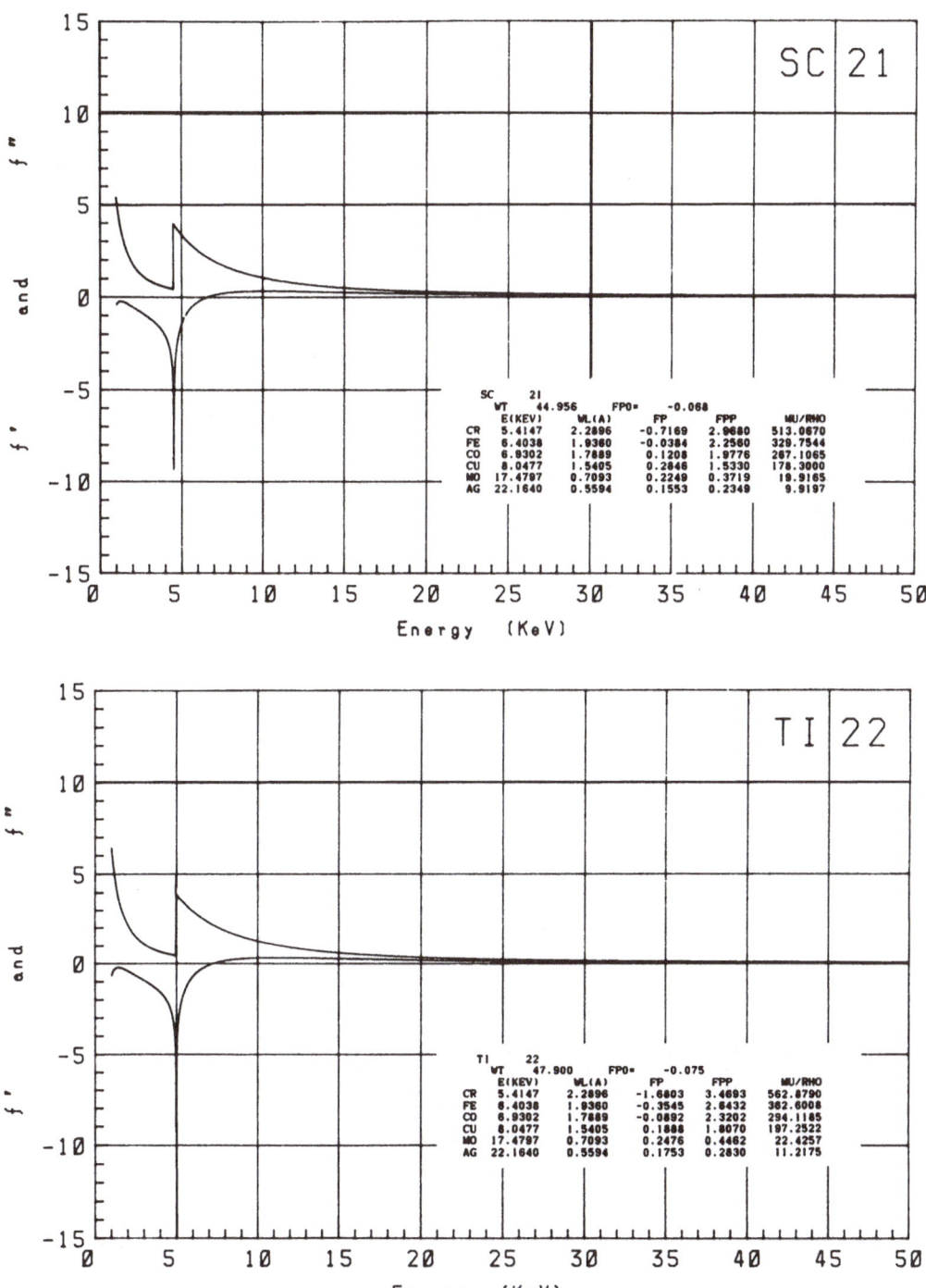

SC 21

SC   21
WT   44.956   FP0=   -0.068

	E(KEV)	WL(A)	FP	FPP	MU/RHO
CR	5.4147	2.2896	-0.7169	2.9680	513.0670
FE	6.4038	1.9360	-0.0384	2.2560	329.7544
CO	6.9302	1.7889	0.1208	1.9776	267.1065
CU	8.0477	1.5405	0.2846	1.5330	178.3000
MO	17.4797	0.7093	0.2249	0.3719	19.9165
AG	22.1640	0.5594	0.1553	0.2349	9.9197

TI 22

TI   22
WT   47.900   FP0=   -0.075

	E(KEV)	WL(A)	FP	FPP	MU/RHO
CR	5.4147	2.2896	-1.6803	3.4693	562.8790
FE	6.4038	1.9360	-0.3545	2.8432	362.6008
CO	6.9302	1.7889	-0.0892	2.3202	294.1185
CU	8.0477	1.5405	0.1888	1.8070	197.2522
MO	17.4797	0.7093	0.2476	0.4462	22.4257
AG	22.1640	0.5594	0.1753	0.2830	11.2175

Energy   (KeV)

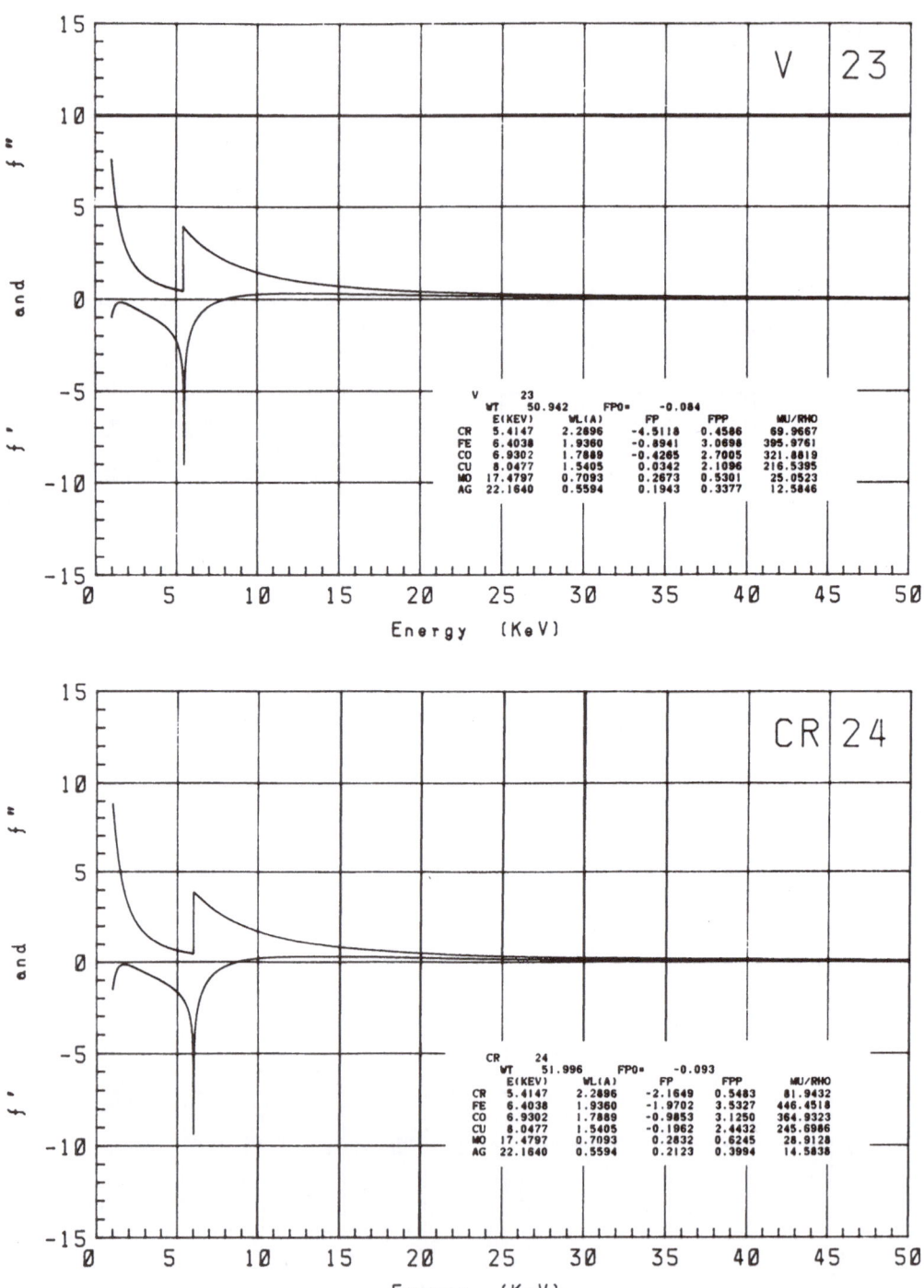

V 23

	E(KEV)	WL(A)	FP	FPP	MU/RHO

V   23
WT   50.942   FPO=   -0.084

	E(KEV)	WL(A)	FP	FPP	MU/RHO
CR	5.4147	2.2896	-4.5118	0.4586	69.9667
FE	6.4038	1.9360	-0.8941	3.0698	395.9761
CO	6.9302	1.7889	-0.4265	2.7005	321.8819
CU	8.0477	1.5405	0.0342	2.1096	216.5395
MO	17.4797	0.7093	0.2673	0.5301	25.0523
AG	22.1640	0.5594	0.1943	0.3377	12.5846

Energy (KeV)

CR 24

CR   24
WT   51.996   FPO=   -0.093

	E(KEV)	WL(A)	FP	FPP	MU/RHO
CR	5.4147	2.2896	-2.1649	0.5483	81.9432
FE	6.4038	1.9360	-1.9702	3.5327	446.4518
CO	6.9302	1.7889	-0.9853	3.1250	364.9323
CU	8.0477	1.5405	-0.1962	2.4432	245.6986
MO	17.4797	0.7093	0.2832	0.6245	28.9128
AG	22.1640	0.5594	0.2123	0.3994	14.5838

Energy (KeV)

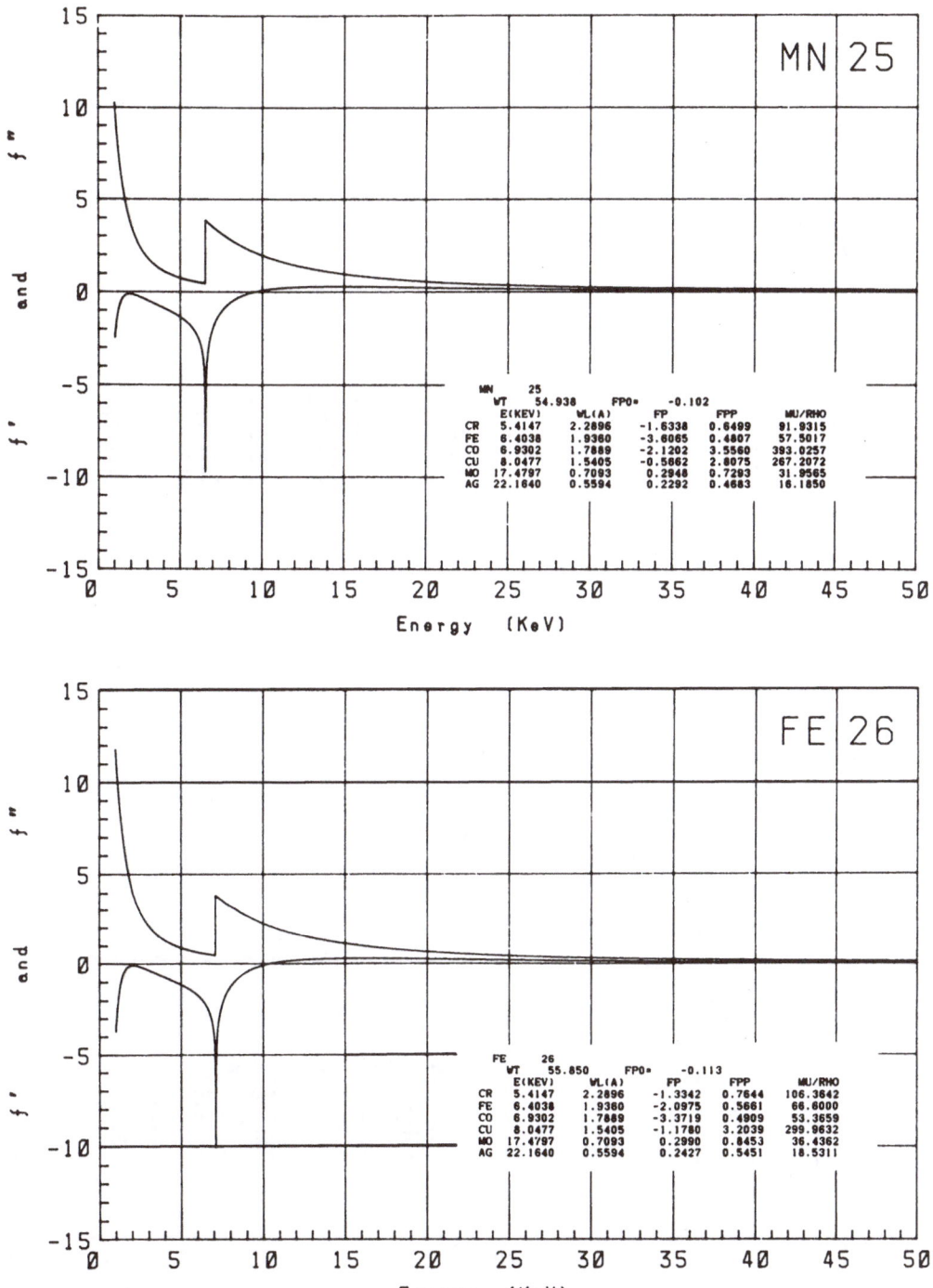

MN 25

	E(KEV)	WL(A)	FP	FPP	MU/RHO
MN 25					
WT 54.938		FPO=	-0.102		
CR	5.4147	2.2896	-1.6338	0.6499	91.9315
FE	6.4038	1.9360	-3.6065	0.4807	57.5017
CO	6.9302	1.7889	-2.1202	3.5560	393.0257
CU	8.0477	1.5405	-0.5862	2.8075	267.2072
MO	17.4797	0.7093	0.2948	0.7293	31.9565
AG	22.1640	0.5594	0.2292	0.4683	16.1850

FE 26

	E(KEV)	WL(A)	FP	FPP	MU/RHO
FE 26					
WT 55.850		FPO=	-0.113		
CR	5.4147	2.2896	-1.3342	0.7644	106.3842
FE	6.4038	1.9360	-2.0975	0.5661	66.6000
CO	6.9302	1.7889	-3.3719	0.4909	53.3659
CU	8.0477	1.5405	-1.1780	3.2039	299.9632
MO	17.4797	0.7093	0.2990	0.8453	36.4362
AG	22.1640	0.5594	0.2427	0.5451	18.5311

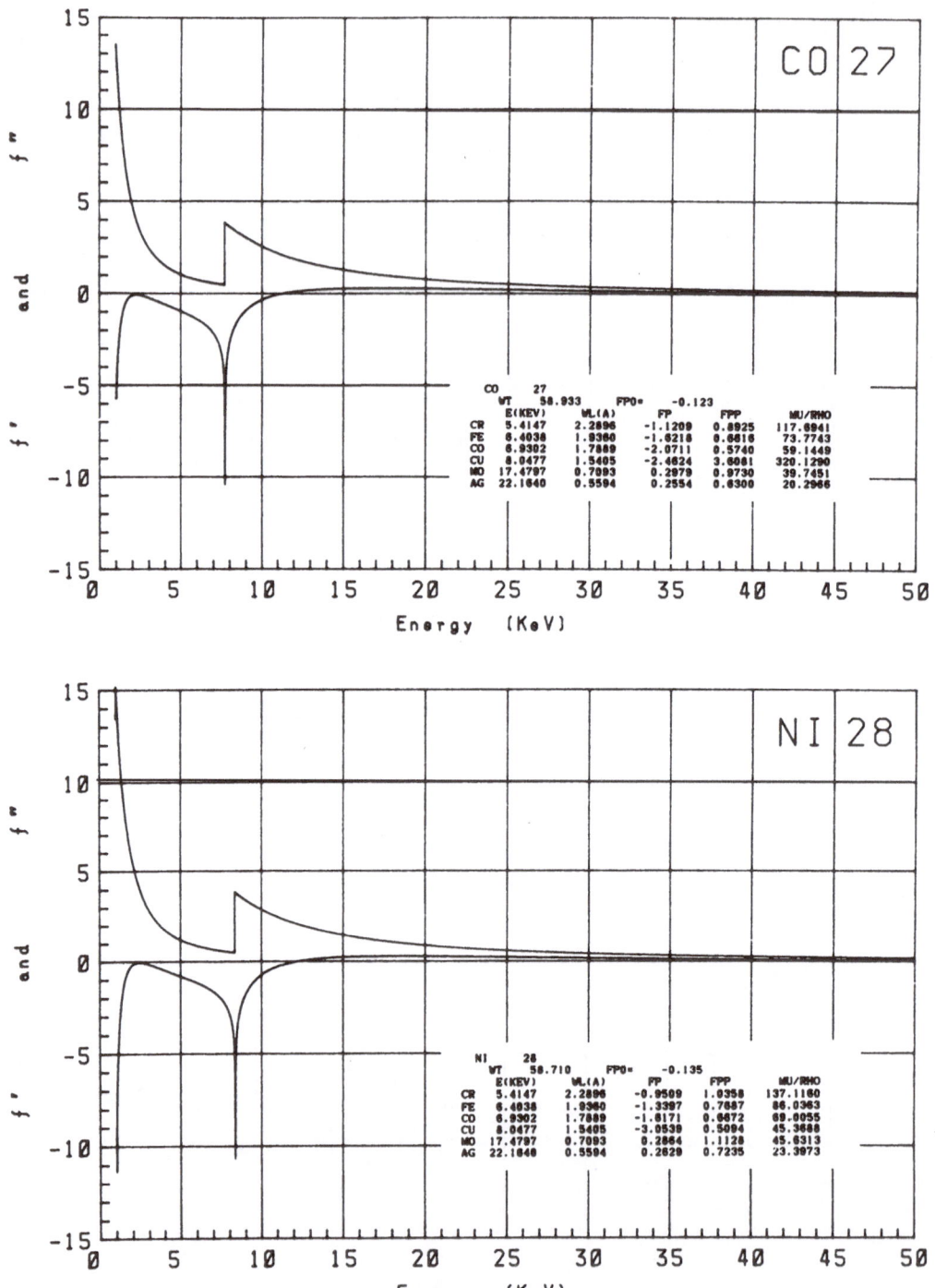

CO 27

CO 27
WT 58.933 FPO= -0.123

	E(KEV)	WL(A)	FP	FPP	MU/RHO
CR	5.4147	2.2896	-1.1209	0.8925	117.6941
FE	6.4038	1.9360	-1.6218	0.6616	73.7743
CO	6.9302	1.7889	-2.0711	0.5740	59.1449
CU	8.0477	1.5405	-2.4624	3.6081	320.1290
MO	17.4797	0.7093	0.2979	0.9730	39.7451
AG	22.1640	0.5594	0.2554	0.6300	20.2966

NI 28

NI 28
WT 58.710 FPO= -0.135

	E(KEV)	WL(A)	FP	FPP	MU/RHO
CR	5.4147	2.2896	-0.9509	1.0358	137.1160
FE	6.4038	1.9360	-1.3397	0.7687	86.0063
CO	6.9302	1.7889	-1.6171	0.6672	69.0055
CU	8.0477	1.5405	-3.0539	0.5094	45.3688
MO	17.4797	0.7093	0.2864	1.1128	45.6313
AG	22.1640	0.5594	0.2629	0.7235	23.3073

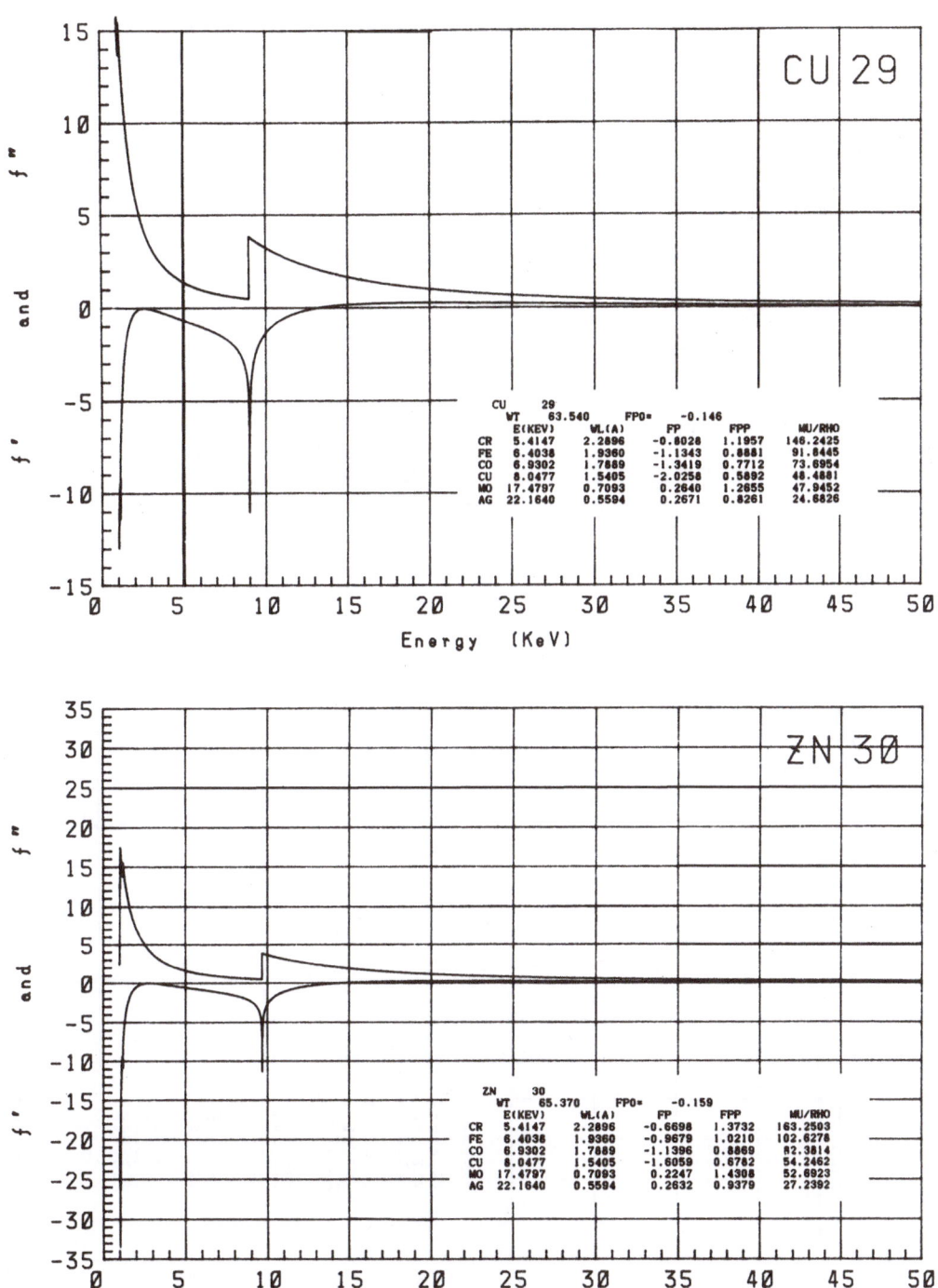

CU 29

CU	29				
WT	63.540	FPO=	-0.146		
	E(KEV)	WL(A)	FP	FPP	MU/RHO
CR	5.4147	2.2896	-0.8028	1.1957	146.2425
FE	6.4038	1.9360	-1.1343	0.8881	91.8445
CO	6.9302	1.7889	-1.3419	0.7712	73.6954
CU	8.0477	1.5405	-2.0258	0.5892	48.4881
MO	17.4797	0.7093	0.2640	1.2655	47.9452
AG	22.1640	0.5594	0.2671	0.8261	24.6826

ZN 30

ZN	30				
WT	65.370	FPO=	-0.159		
	E(KEV)	WL(A)	FP	FPP	MU/RHO
CR	5.4147	2.2896	-0.6698	1.3732	163.2503
FE	6.4038	1.9360	-0.9679	1.0210	102.6278
CO	6.9302	1.7889	-1.1396	0.8869	82.3814
CU	8.0477	1.5405	-1.6059	0.6782	54.2462
MO	17.4797	0.7093	0.2247	1.4308	52.6923
AG	22.1640	0.5594	0.2632	0.9379	27.2392

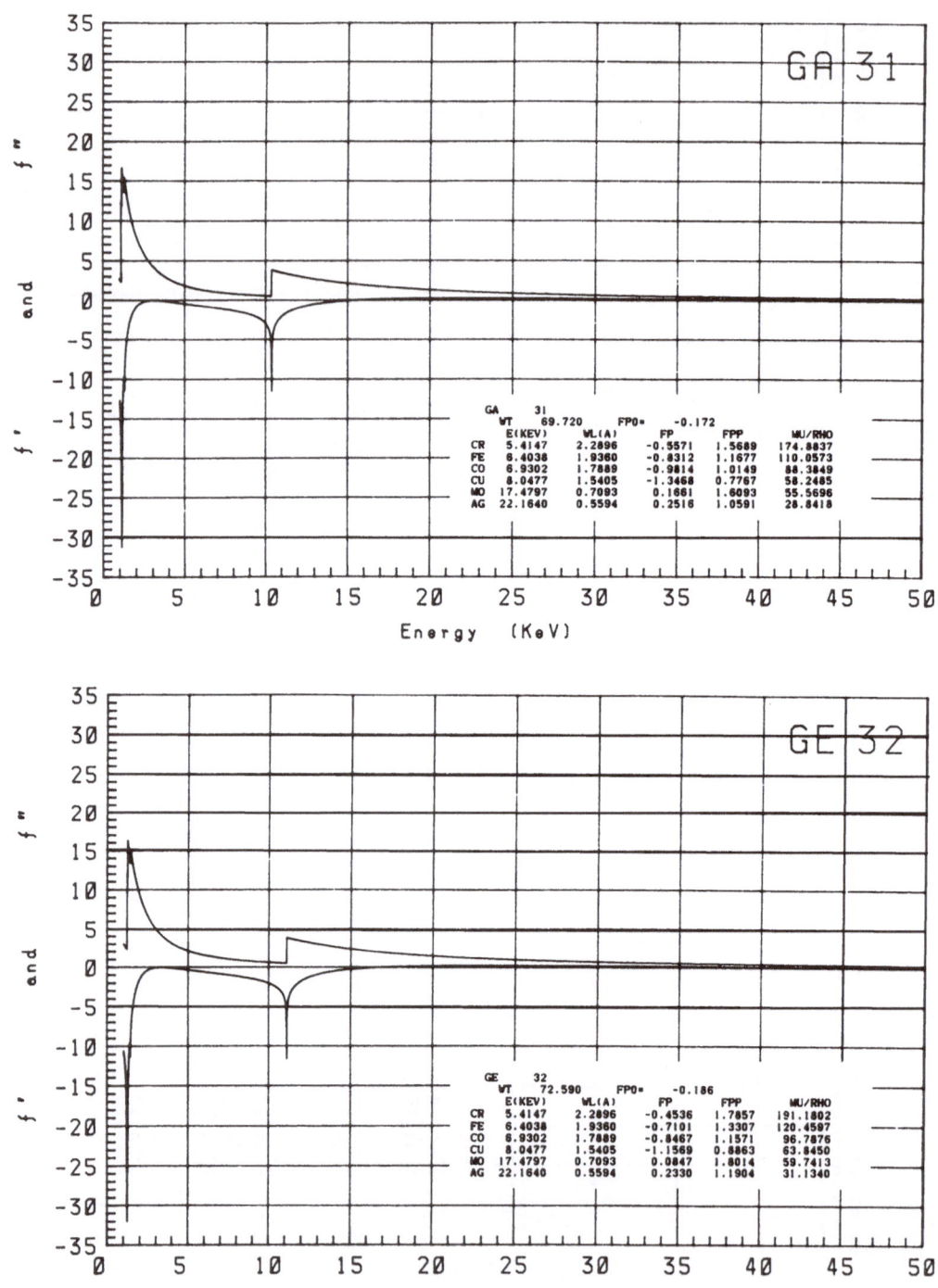

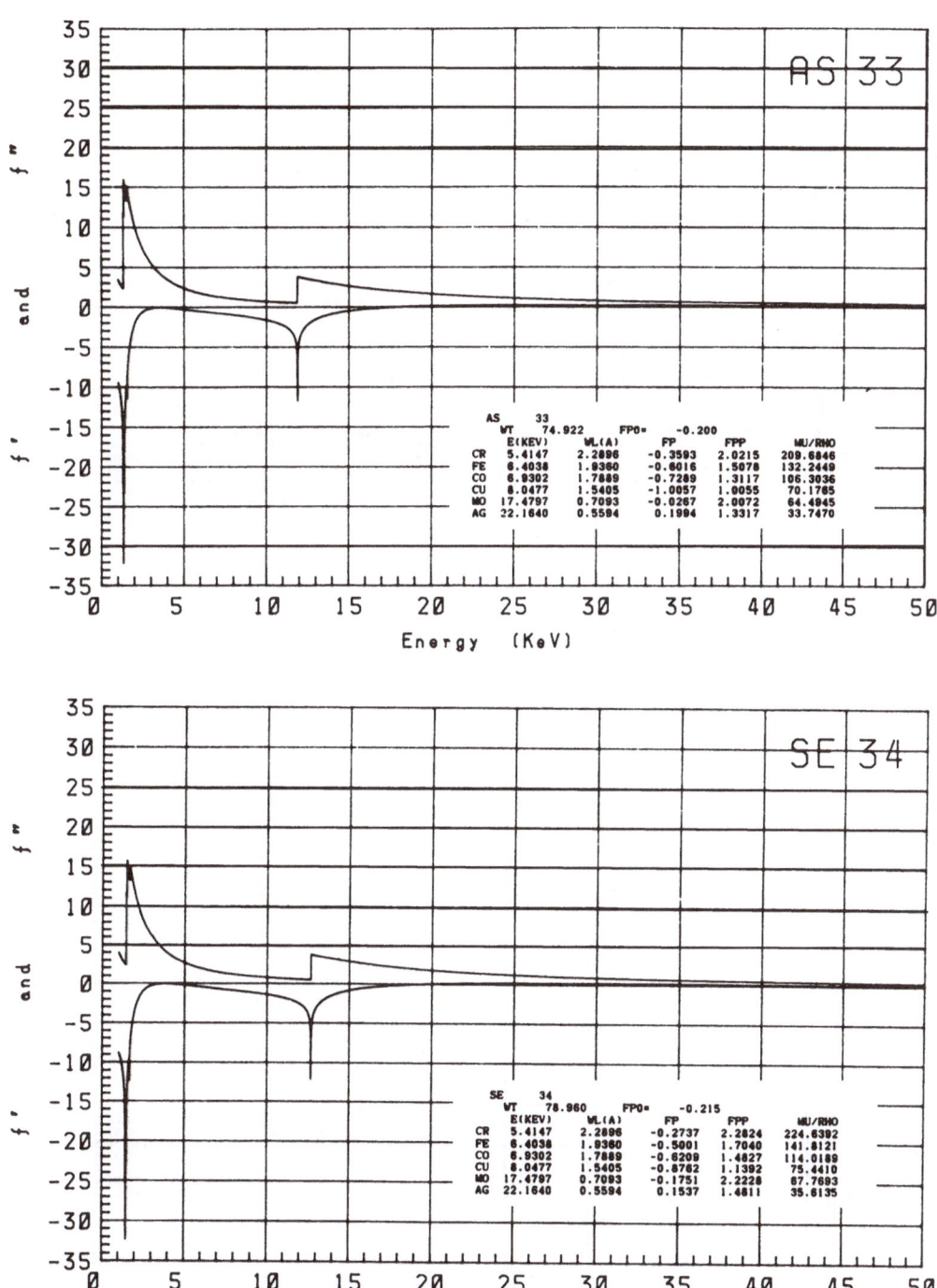

AS 33

AS	33				
	WT	74.922	FPO=	-0.200	
	E(KEV)	WL(A)	FP	FPP	MU/RHO
CR	5.4147	2.2896	-0.3593	2.0215	209.6846
FE	6.4038	1.9360	-0.6016	1.5078	132.2449
CO	6.9302	1.7889	-0.7289	1.3117	106.3036
CU	8.0477	1.5405	-1.0057	1.0055	70.1785
MO	17.4797	0.7093	-0.0267	2.0072	64.4945
AG	22.1640	0.5594	0.1994	1.3317	33.7470

SE 34

SE	34				
	WT	78.960	FPO=	-0.215	
	E(KEV)	WL(A)	FP	FPP	MU/RHO
CR	5.4147	2.2896	-0.2737	2.2824	224.6392
FE	6.4038	1.9360	-0.5001	1.7040	141.8121
CO	6.9302	1.7889	-0.6209	1.4827	114.0169
CU	8.0477	1.5405	-0.8762	1.1392	75.4410
MO	17.4797	0.7093	-0.1751	2.2228	87.7693
AG	22.1640	0.5594	0.1537	1.4811	35.6135

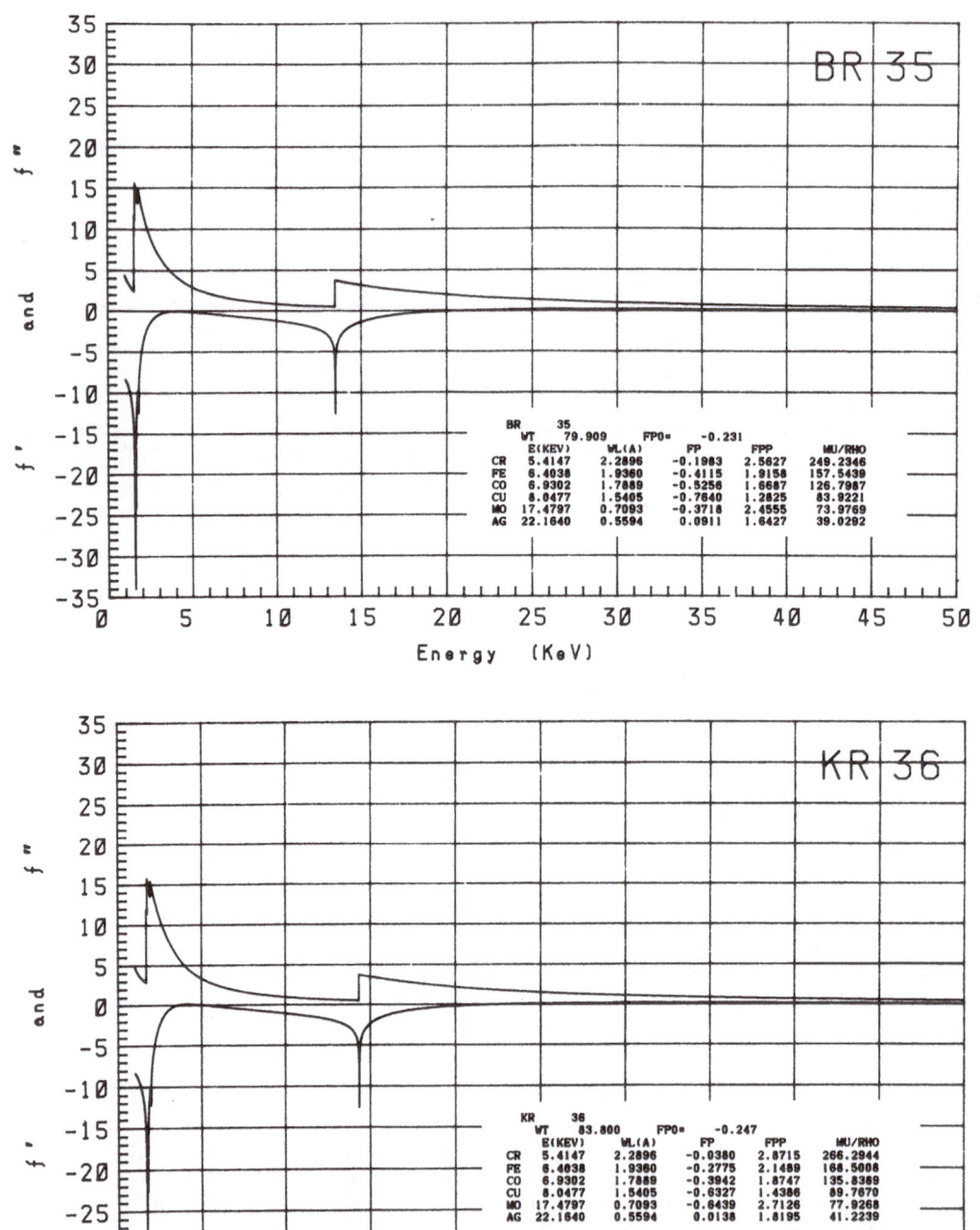

BR 35

BR   35
WT   79.909   FPO=   -0.231
	E(KEV)	WL(A)	FP	FPP	MU/RHO
CR	5.4147	2.2896	-0.1963	2.5627	249.2346
FE	6.4038	1.9360	-0.4115	1.9158	157.5439
CO	6.9302	1.7889	-0.5256	1.6687	126.7987
CU	8.0477	1.5405	-0.7840	1.2825	83.9221
MO	17.4797	0.7093	-0.3718	2.4555	73.9769
AG	22.1640	0.5594	0.0911	1.6427	39.0292

KR 36

KR   36
WT   83.800   FPO=   -0.247
	E(KEV)	WL(A)	FP	FPP	MU/RHO
CR	5.4147	2.2896	-0.0380	2.8715	266.2944
FE	6.4038	1.9360	-0.2775	2.1489	168.5008
CO	6.9302	1.7889	-0.3942	1.8747	135.8389
CU	8.0477	1.5405	-0.6327	1.4386	89.7670
MO	17.4797	0.7093	-0.6439	2.7126	77.9268
AG	22.1640	0.5594	0.0138	1.8195	41.2239

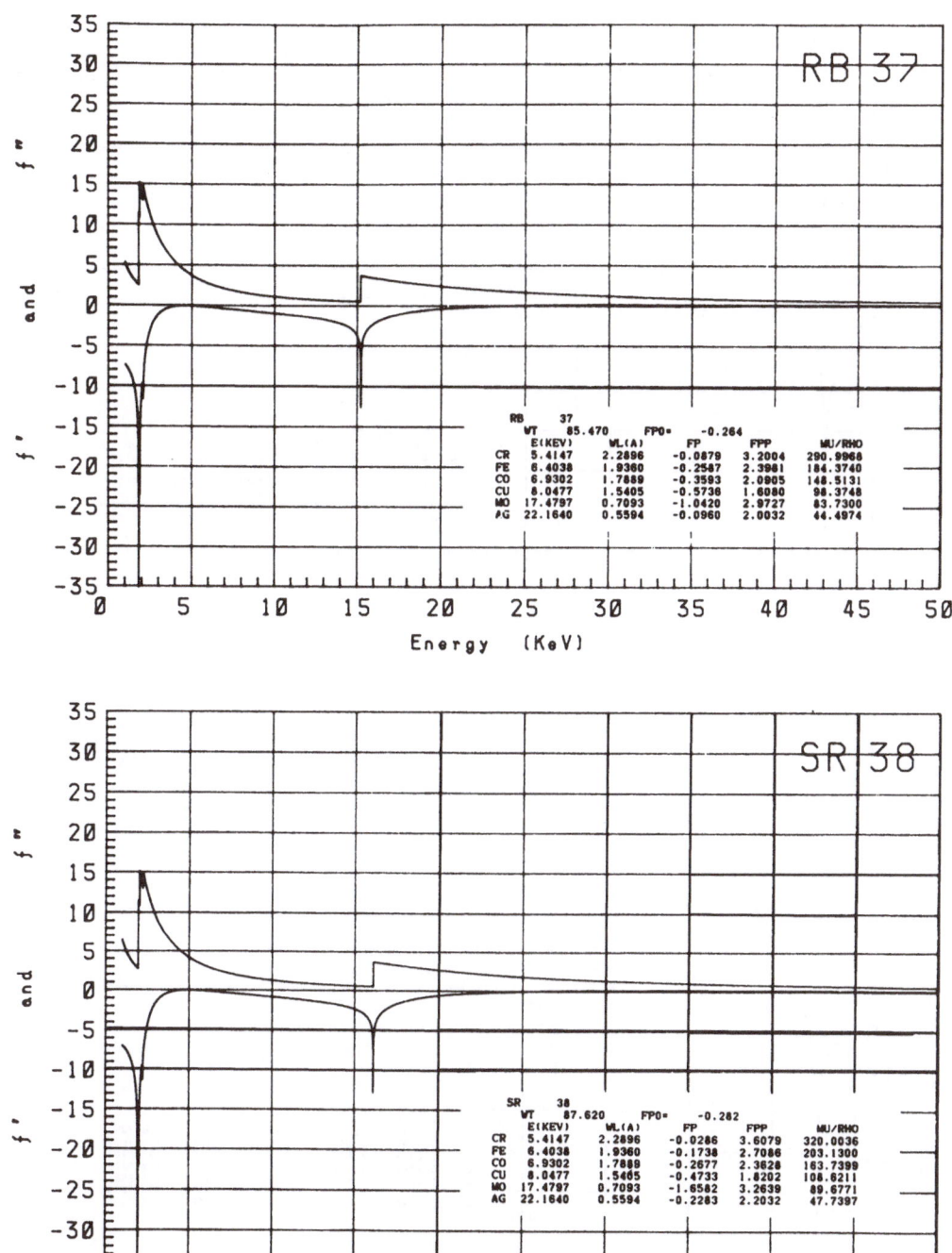

RB   37
                        WT   85.470   FPO=   -0.264
          E(KEV)   WL(A)   FP   FPP   MU/RHO
CR   5.4147   2.2896   -0.0879   3.2004   290.9968
FE   6.4038   1.9360   -0.2587   2.3981   184.3740
CO   6.9302   1.7889   -0.3593   2.0905   148.5131
CU   8.0477   1.5405   -0.5736   1.6080   98.3748
MO   17.4797   0.7093   -1.0420   2.9727   83.7300
AG   22.1640   0.5594   -0.0960   2.0032   44.4974

SR   38
                        WT   87.620   FPO=   -0.282
          E(KEV)   WL(A)   FP   FPP   MU/RHO
CR   5.4147   2.2896   -0.0286   3.6079   320.0036
FE   6.4038   1.9360   -0.1738   2.7086   203.1300
CO   6.9302   1.7889   -0.2677   2.3828   163.7399
CU   8.0477   1.5465   -0.4733   1.8202   108.6211
MO   17.4797   0.7093   -1.6582   3.2639   89.6771
AG   22.1640   0.5594   -0.2283   2.2032   47.7397

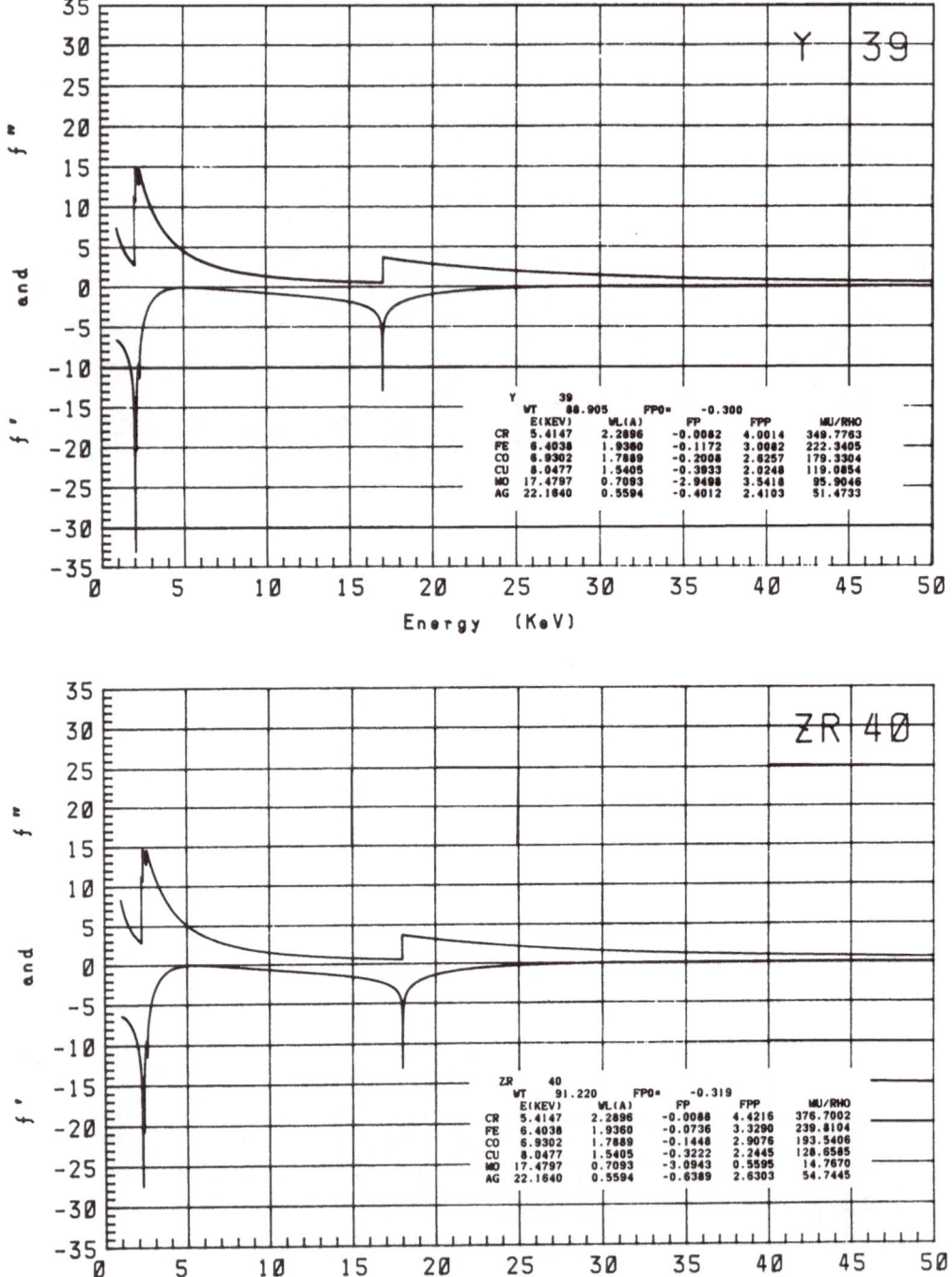

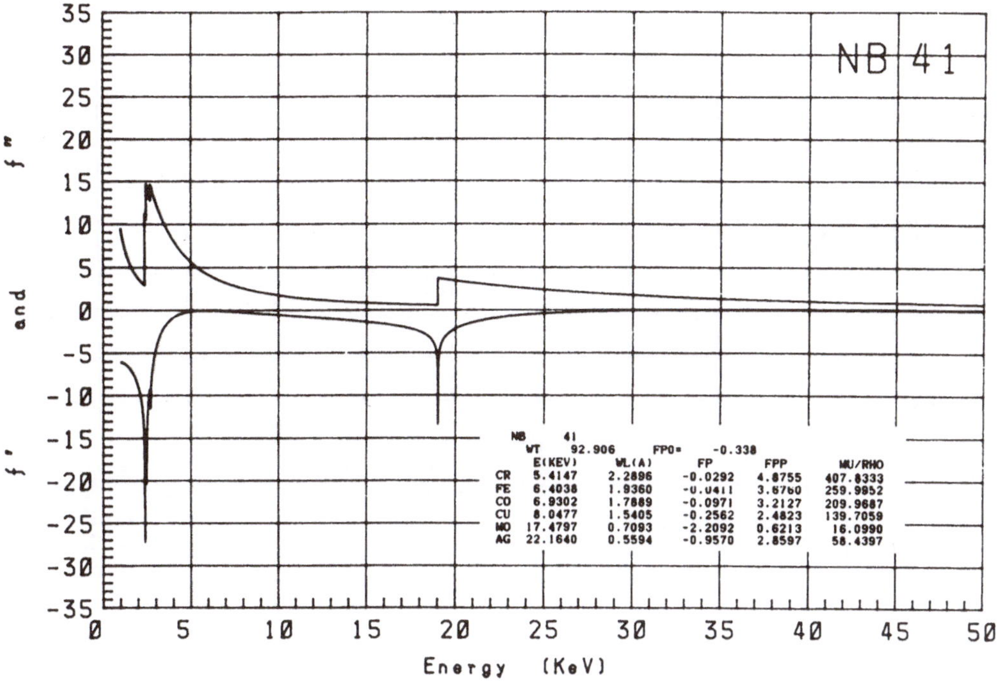

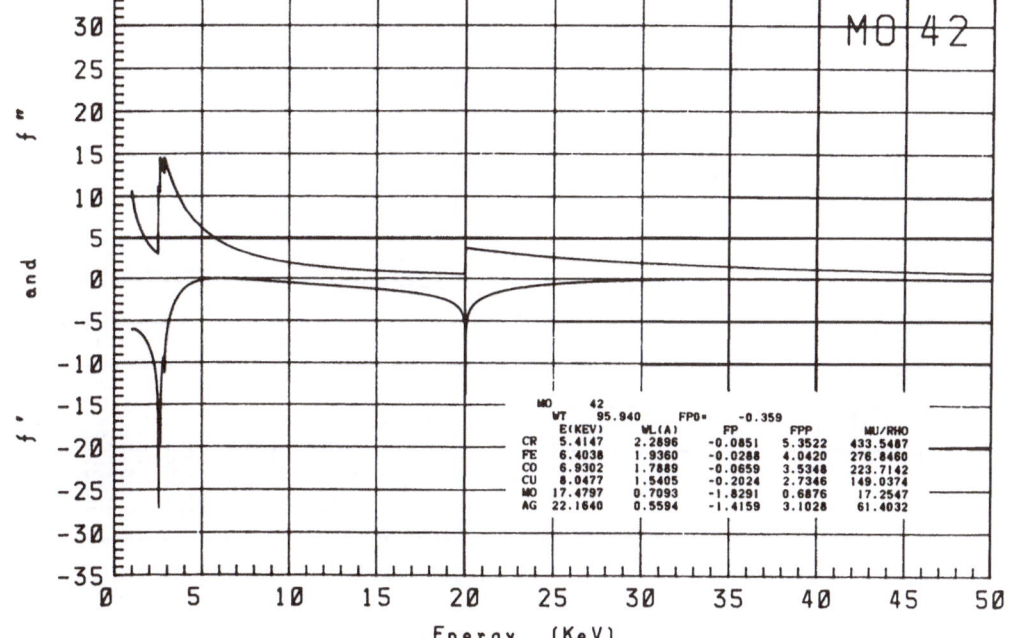

NB 41

		NB	41		
		WT	92.906	FPO=	-0.338
	E(KEV)	WL(A)	FP	FPP	MU/RHO
CR	5.4147	2.2896	-0.0292	4.8755	407.8333
FE	6.4038	1.9360	-0.0411	3.8760	259.9852
CO	6.9302	1.7889	-0.0971	3.2127	209.9687
CU	8.0477	1.5405	-0.2562	2.4823	139.7059
MO	17.4797	0.7093	-2.2092	0.6213	16.0990
AG	22.1640	0.5594	-0.9570	2.8597	58.4397

MO 42

		MO	42		
		WT	95.940	FPO=	-0.359
	E(KEV)	WL(A)	FP	FPP	MU/RHO
CR	5.4147	2.2896	-0.0851	5.3522	433.5487
FE	6.4038	1.9360	-0.0288	4.0420	276.8460
CO	6.9302	1.7889	-0.0659	3.5348	223.7142
CU	8.0477	1.5405	-0.2024	2.7346	149.0374
MO	17.4797	0.7093	-1.8291	0.6876	17.2547
AG	22.1640	0.5594	-1.4159	3.1028	61.4032

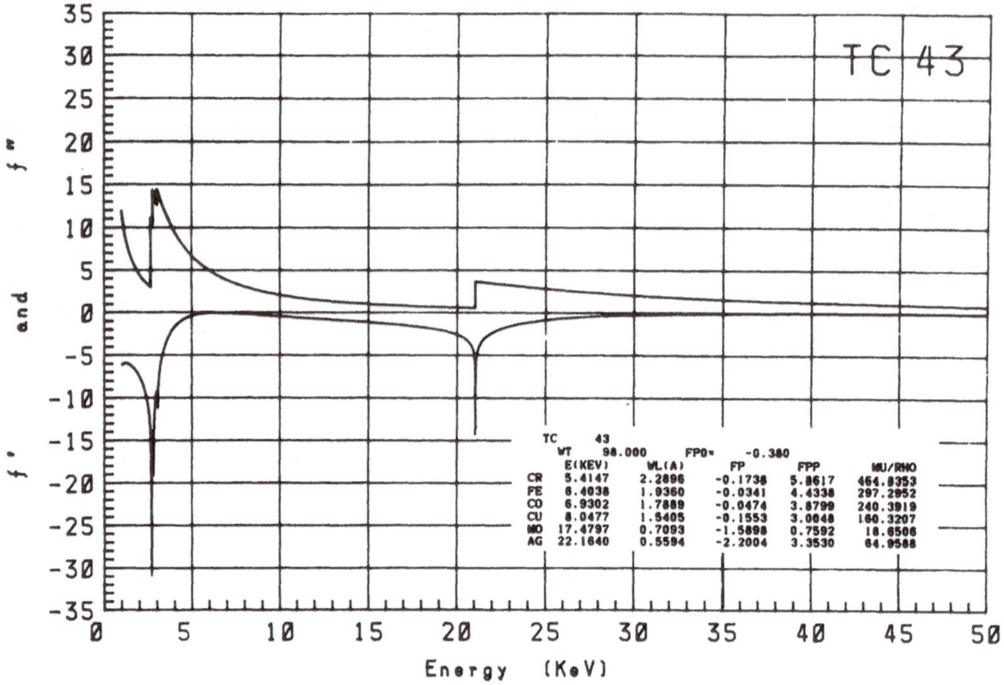

TC 43

TC	43				
	WT	98.000	FP0=	-0.380	
	E(KEV)	WL(A)	FP	FPP	MU/RHO
CR	5.4147	2.2896	-0.1738	5.8617	464.8353
FE	6.4038	1.9360	-0.0341	4.4338	297.2952
CO	6.9302	1.7889	-0.0474	3.8799	240.3819
CU	8.0477	1.5405	-0.1553	3.0848	160.3207
MO	17.4797	0.7093	-1.5898	0.7592	18.8506
AG	22.1640	0.5594	-2.2004	3.3530	64.9588

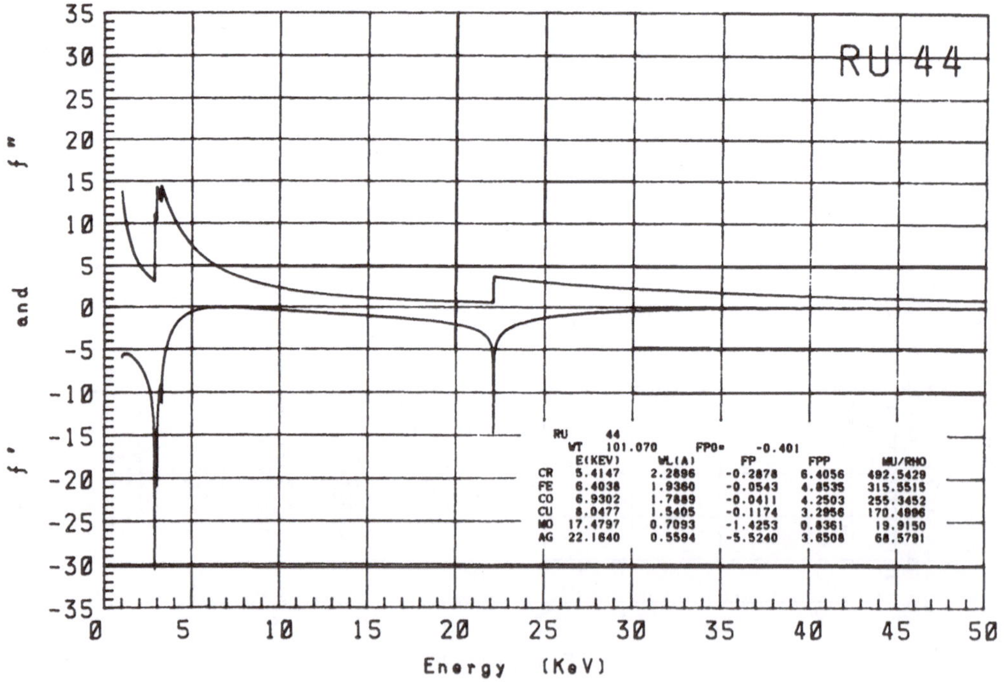

RU 44

RU	44				
	WT	101.070	FP0=	-0.401	
	E(KEV)	WL(A)	FP	FPP	MU/RHO
CR	5.4147	2.2896	-0.2878	6.4056	492.5429
FE	6.4038	1.9360	-0.0543	4.8535	315.5515
CO	6.9302	1.7889	-0.0411	4.2503	255.3452
CU	8.0477	1.5405	-0.1174	3.2956	170.4996
MO	17.4797	0.7093	-1.4253	0.8361	19.9150
AG	22.1640	0.5594	-5.5240	3.6508	68.5791

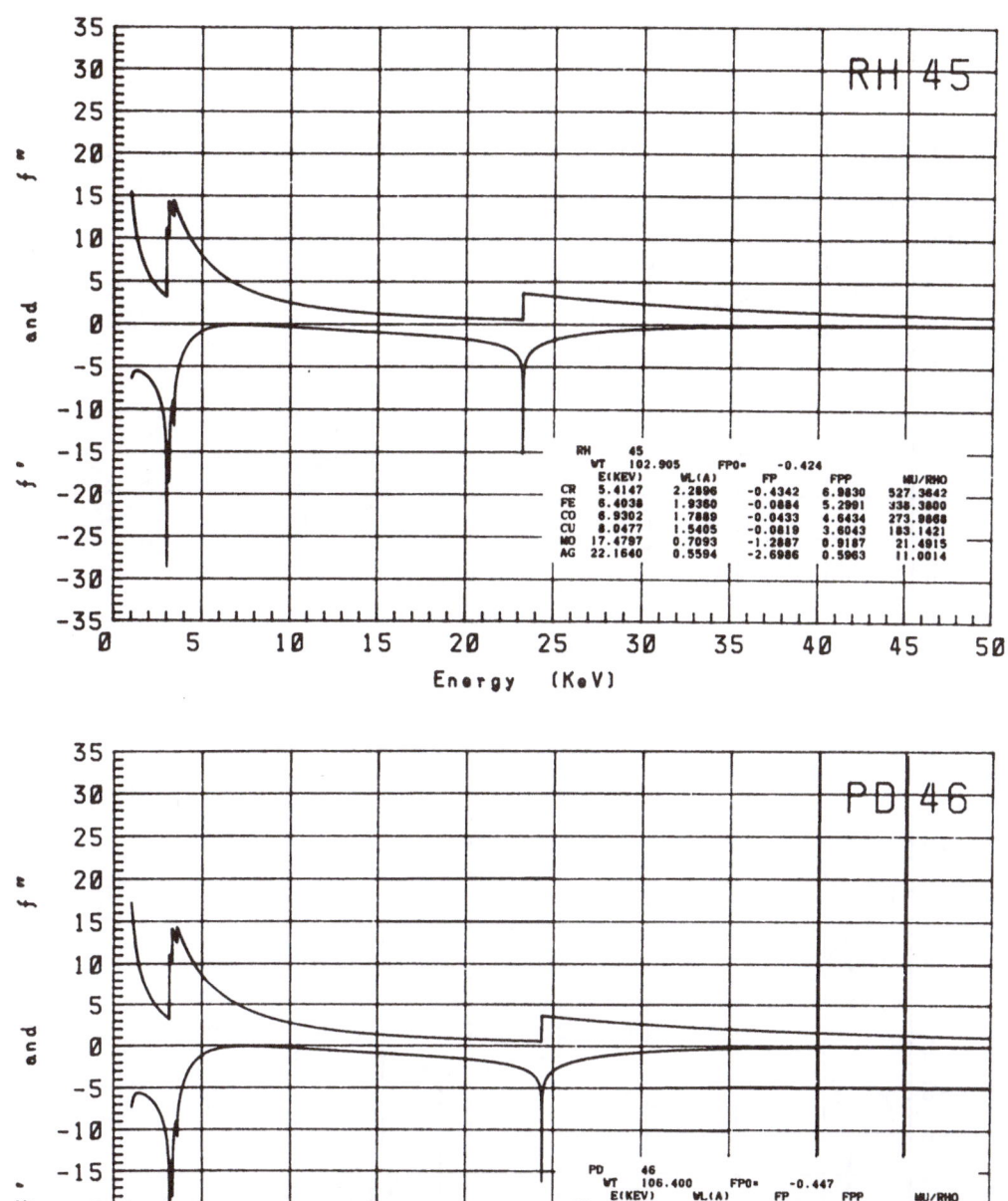

RH 45

RH	45				
	WT	102.905	FPO=	-0.424	
	E(KEV)	WL(A)	FP	FPP	MU/RHO
CR	5.4147	2.2896	-0.4342	6.9830	527.3642
FE	6.4038	1.9360	-0.0884	5.2991	338.3800
CO	6.9302	1.7889	-0.0433	4.6434	273.9668
CU	8.0477	1.5405	-0.0819	3.6043	183.1421
MO	17.4797	0.7093	-1.2887	0.9187	21.4915
AG	22.1640	0.5594	-2.6986	0.5963	11.0014

PD 46

PD	46				
	WT	106.400	FPO=	-0.447	
	E(KEV)	WL(A)	FP	FPP	MU/RHO
CR	5.4147	2.2896	-0.6637	7.5936	554.6357
FE	6.4038	1.9360	-0.1714	5.7719	356.4651
CO	6.9302	1.7889	-0.0848	5.0613	288.8351
CU	8.0477	1.5405	-0.0725	3.9335	193.3079
MO	17.4797	0.7093	-1.1820	1.0070	22.7841
AG	22.1640	0.5594	-2.1374	0.6544	11.6763

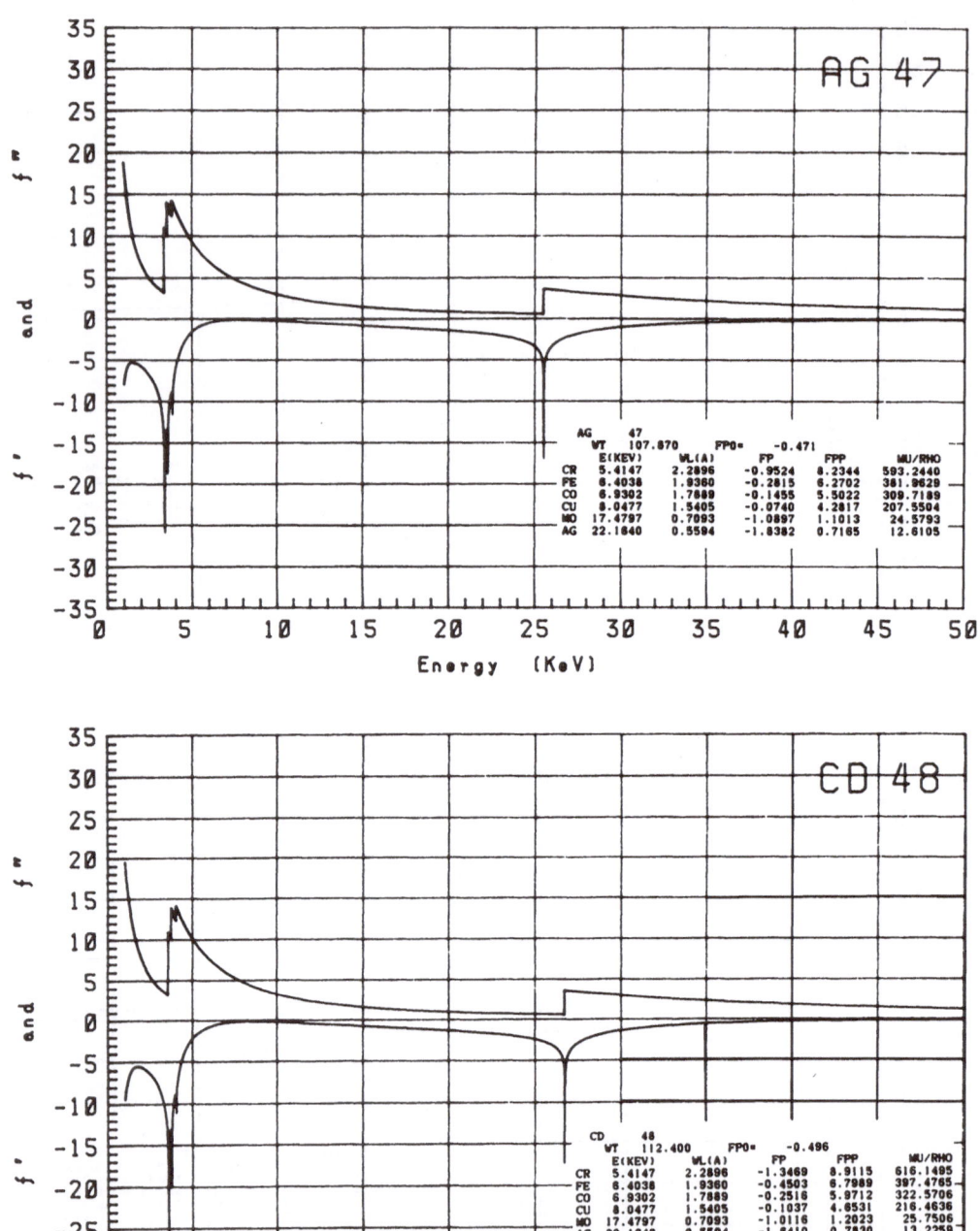

AG 47

AG   47
WT   107.870   FPO=   -0.471
	E(KEV)	WL(A)	FP	FPP	MU/RHO
CR	5.4147	2.2896	-0.9524	8.2344	593.2440
FE	8.4038	1.9360	-0.2815	6.2702	381.9629
CO	6.9302	1.7889	-0.1455	5.5022	309.7189
CU	8.0477	1.5405	-0.0740	4.2817	207.5504
MO	17.4797	0.7093	-1.0897	1.1013	24.5793
AG	22.1640	0.5594	-1.8382	0.7165	12.6105

Energy (KeV)

CD 48

CD   48
WT   112.400   FPO=   -0.496
	E(KEV)	WL(A)	FP	FPP	MU/RHO
CR	5.4147	2.2896	-1.3469	8.9115	616.1495
FE	8.4038	1.9360	-0.4503	6.7989	397.4765
CO	6.9302	1.7889	-0.2516	5.9712	322.5706
CU	8.0477	1.5405	-0.1037	4.6531	216.4636
MO	17.4797	0.7093	-1.0116	1.2023	25.7506
AG	22.1640	0.5594	-1.6410	0.7830	13.2258

Energy (KeV)

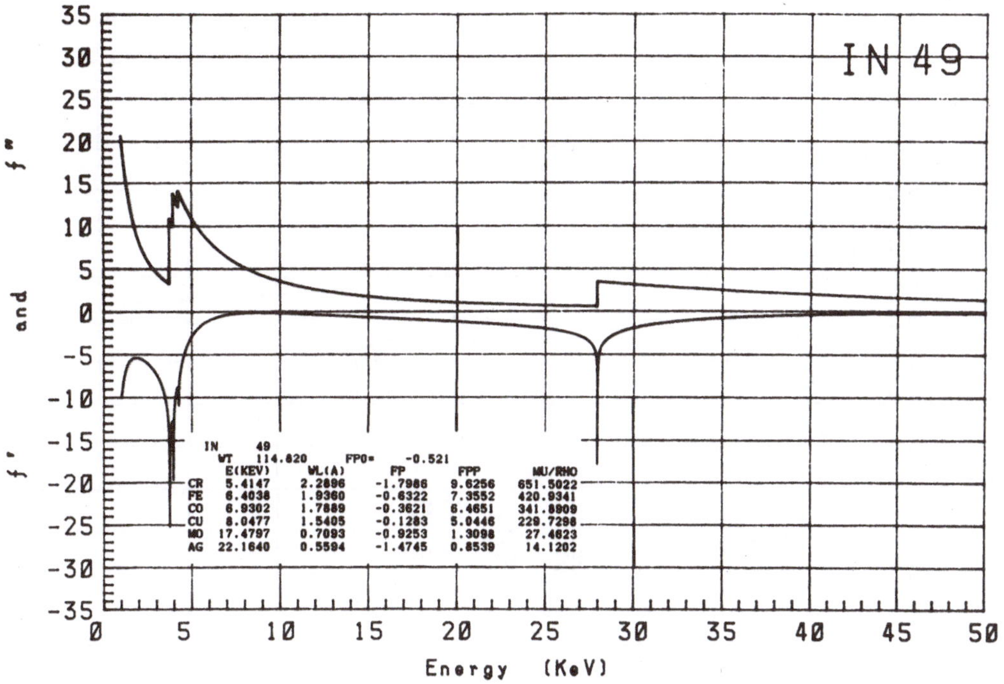

IN 49

IN 49					
WT	114.820	FPO=	-0.521		
	E(KEV)	WL(A)	FP	FPP	MU/RHO
CR	5.4147	2.2896	-1.7986	9.6256	651.5022
FE	6.4038	1.9360	-0.6322	7.3552	420.8941
CO	6.9302	1.7889	-0.3621	6.4651	341.8909
CU	8.0477	1.5405	-0.1283	5.0446	229.7298
MO	17.4797	0.7093	-0.9253	1.3098	27.4623
AG	22.1640	0.5594	-1.4745	0.8539	14.1202

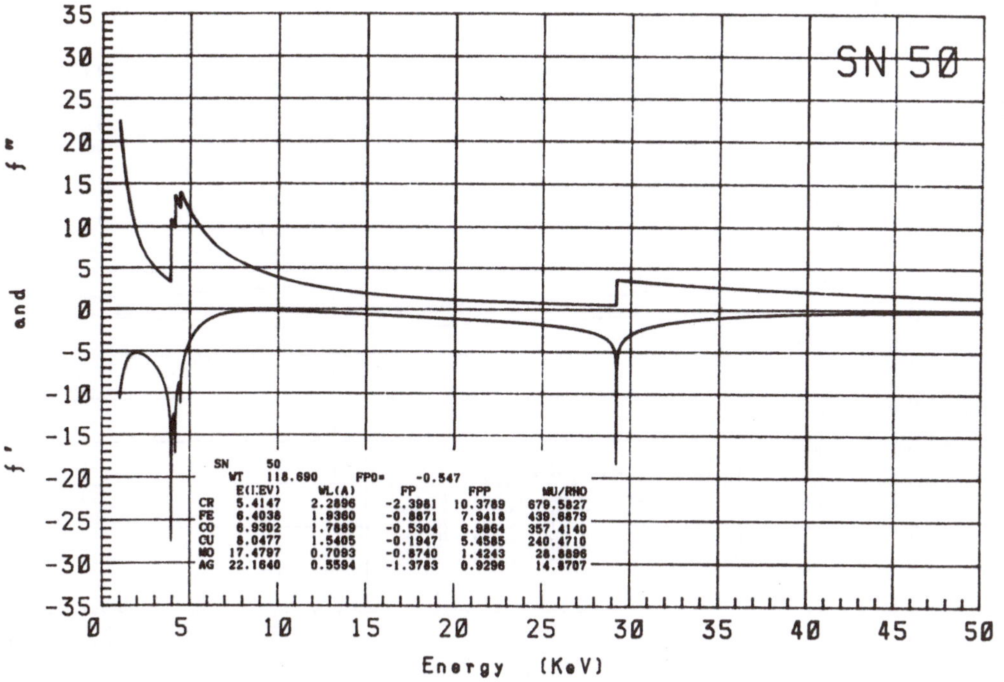

SN 50

SN 50					
WT	118.690	FPO=	-0.547		
	E(KEV)	WL(A)	FP	FPP	MU/RHO
CR	5.4147	2.2896	-2.3981	10.3789	679.5827
FE	6.4038	1.9360	-0.8871	7.9418	439.6879
CO	6.9302	1.7889	-0.5304	6.9864	357.4140
CU	8.0477	1.5405	-0.1947	5.4585	240.4710
MO	17.4797	0.7093	-0.8740	1.4243	26.8896
AG	22.1640	0.5594	-1.3783	0.9296	14.8707

SB 51
f'' and f'
SB 51
WT 121.750 FPO= -0.575
E(KEV) WL(A) FP FPP MU/RHO
CR 5.4147 2.2896 -3.1778 11.1646 712.6534
FE 6.4038 1.9360 -1.2031 8.5563 461.8023
CO 6.9302 1.7889 -0.7405 7.5321 375.6456
CU 8.0477 1.5405 -0.2822 5.8935 253.1129
MO 17.4797 0.7093 -0.8167 1.5459 30.5667
AG 22.1640 0.5594 -1.2848 1.0101 15.7521
Energy (KeV)

TE 52
f'' and f'
TE 52
WT 127.600 FPO= -0.602
E(KEV) WL(A) FP FPP MU/RHO
CR 5.4147 2.2896 -4.2462 11.9943 730.5117
FE 6.4038 1.9360 -1.6108 9.2023 473.8997
CO 6.9302 1.7889 -1.0148 8.1052 385.6933
CU 8.0477 1.5405 -0.4066 6.3518 260.2868
MO 17.4797 0.7093 -0.7646 1.6749 31.5993
AG 22.1640 0.5594 -1.2050 1.0957 16.3033
Energy (KeV)

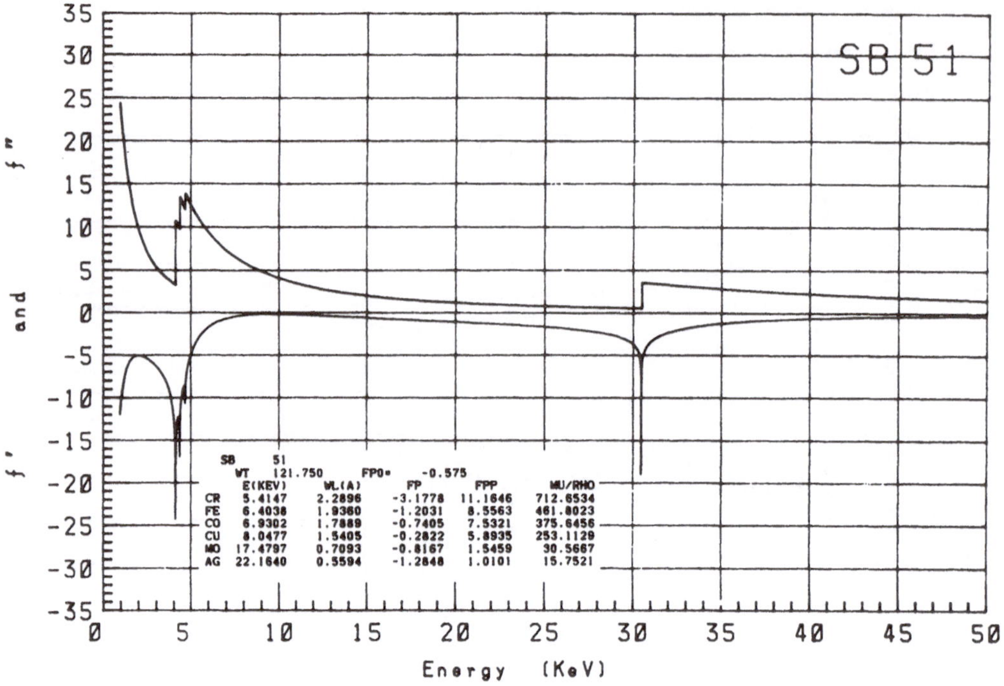

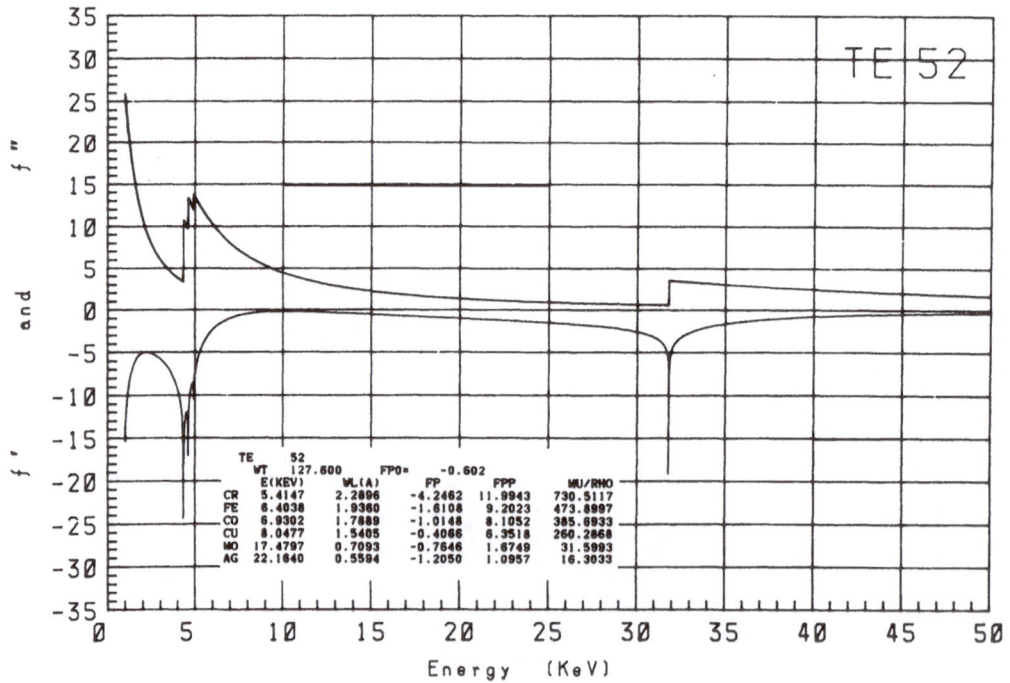

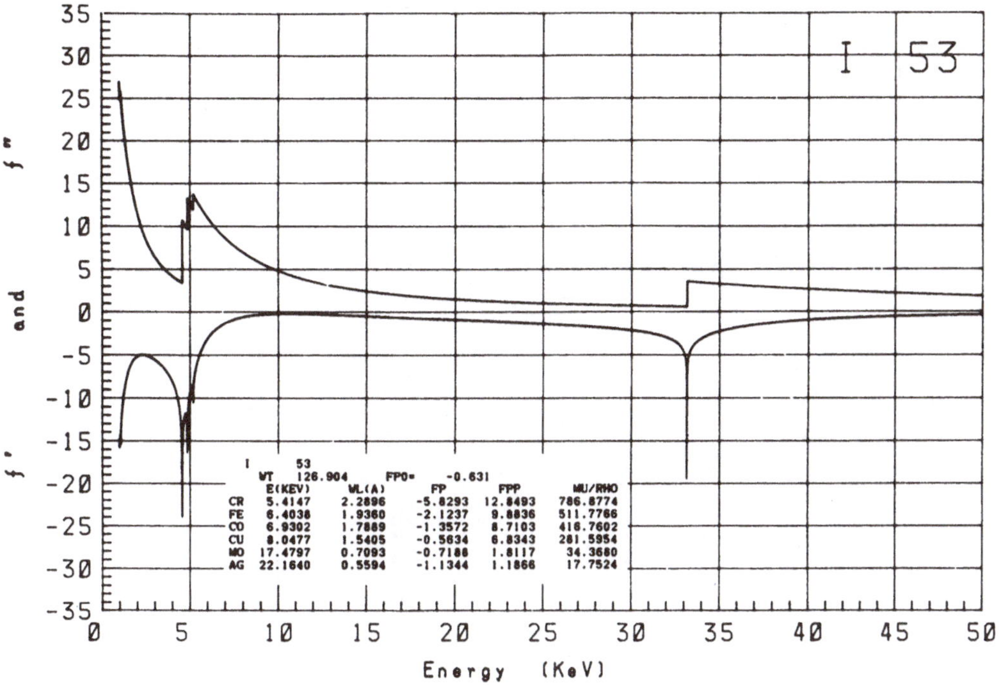
I 53
f'' and f'
Energy (KeV)
0 5 10 15 20 25 30 35 40 45 50
35 30 25 20 15 10 5 0 -5 -10 -15 -20 -25 -30 -35
I 53
WT 126.904 FPO= -0.631
E(KEV) WL(A) FP FPP MU/RHO
CR 5.4147 2.2896 -5.8293 12.8493 786.8774
FE 6.4038 1.9360 -2.1237 9.8836 511.7766
CO 6.9302 1.7889 -1.3572 8.7103 416.7602
CU 8.0477 1.5405 -0.5634 6.8343 281.5954
MO 17.4797 0.7093 -0.7188 1.8117 34.3680
AG 22.1640 0.5594 -1.1344 1.1866 17.7524

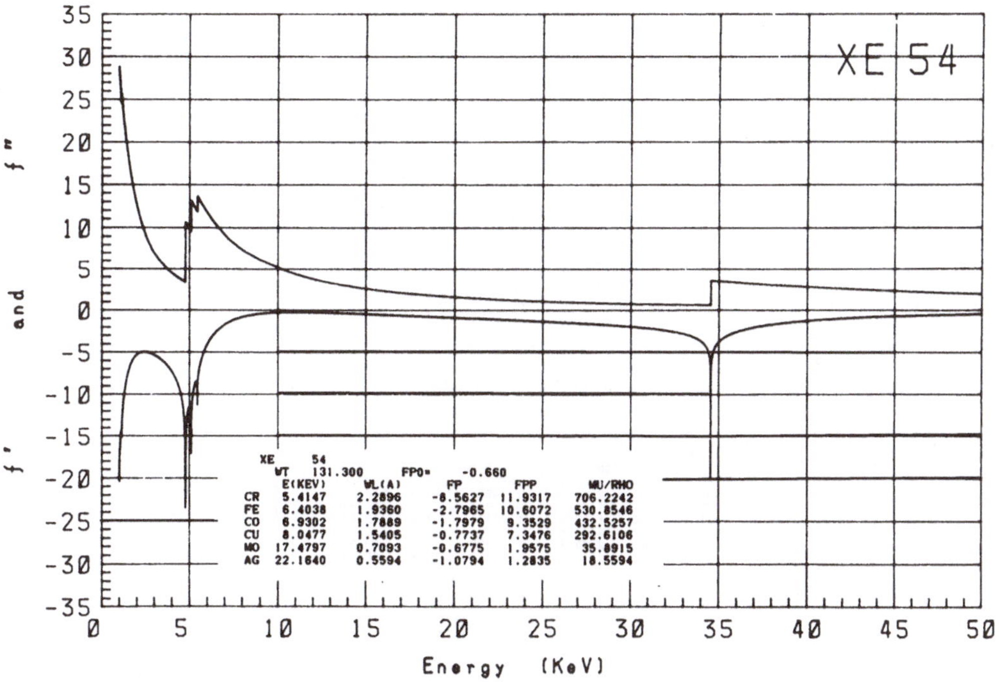
XE 54
f'' and f'
Energy (KeV)
0 5 10 15 20 25 30 35 40 45 50
35 30 25 20 15 10 5 0 -5 -10 -15 -20 -25 -30 -35
XE 54
WT 131.300 FPO= -0.660
E(KEV) WL(A) FP FPP MU/RHO
CR 5.4147 2.2896 -8.5627 11.9317 706.2242
FE 6.4038 1.9360 -2.7965 10.6072 530.8546
CO 6.9302 1.7889 -1.7979 9.3529 432.5257
CU 8.0477 1.5405 -0.7737 7.3476 292.6106
MO 17.4797 0.7093 -0.6775 1.9575 35.8915
AG 22.1640 0.5594 -1.0794 1.2835 18.5594

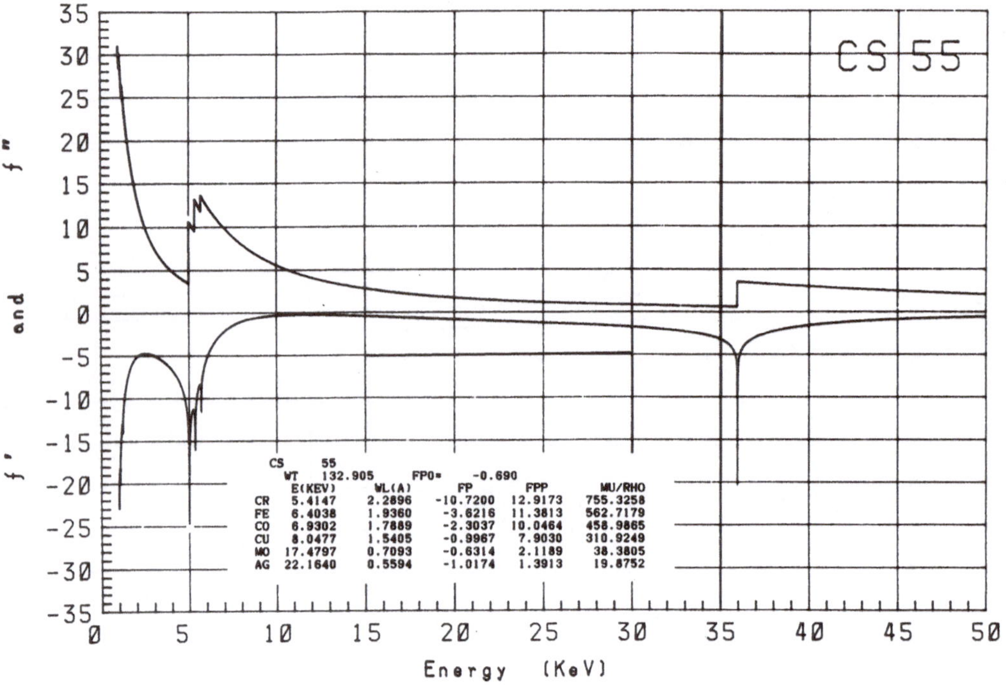

CS 55

CS	55	WT	132.905	FPO=	-0.690		
	E(KEV)	WL(A)	FP	FPP	MU/RHO		
CR	5.4147	2.2896	-10.7200	12.9173	755.3258		
FE	6.4038	1.9360	-3.6216	11.3813	562.7179		
CO	6.9302	1.7889	-2.3037	10.0464	458.9865		
CU	8.0477	1.5405	-0.9967	7.9030	310.9249		
MO	17.4797	0.7093	-0.6314	2.1189	38.3805		
AG	22.1640	0.5594	-1.0174	1.3913	19.8752		

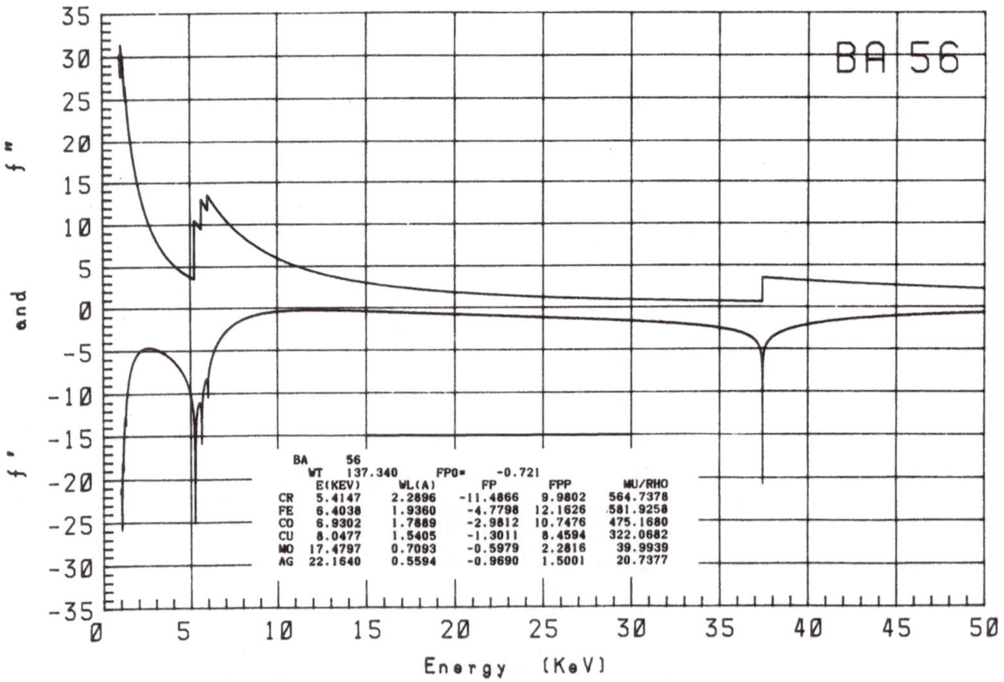

BA 56

BA	56	WT	137.340	FPO=	-0.721		
	E(KEV)	WL(A)	FP	FPP	MU/RHO		
CR	5.4147	2.2896	-11.4866	9.9802	564.7378		
FE	6.4038	1.9360	-4.7798	12.1626	581.9258		
CO	6.9302	1.7889	-2.9812	10.7476	475.1680		
CU	8.0477	1.5405	-1.3011	8.4594	322.0682		
MO	17.4797	0.7093	-0.5979	2.2816	39.9939		
AG	22.1640	0.5594	-0.9690	1.5001	20.7377		

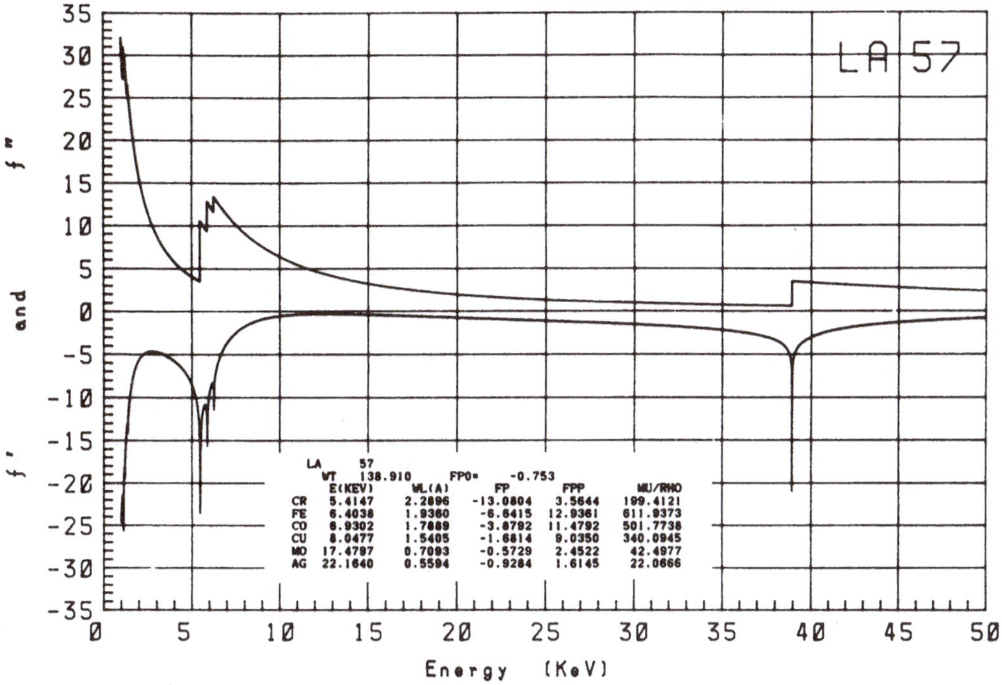
LA 57
f″ and f′
35
30
25
20
15
10
5
0
-5
-10
-15
-20
-25
-30
-35
0 5 10 15 20 25 30 35 40 45 50
Energy (KeV)
LA 57
WT 138.910 FPO= -0.753
E(KEV) WL(A) FP FPP MU/RHO
CR 5.4147 2.2896 -13.0804 3.5644 199.4121
FE 6.4038 1.9360 -6.6415 12.9361 611.9373
CO 6.9302 1.7889 -3.8792 11.4792 501.7738
CU 8.0477 1.5405 -1.6814 9.0350 340.0945
MO 17.4797 0.7093 -0.5729 2.4522 42.4977
AG 22.1640 0.5594 -0.9284 1.6145 22.0666

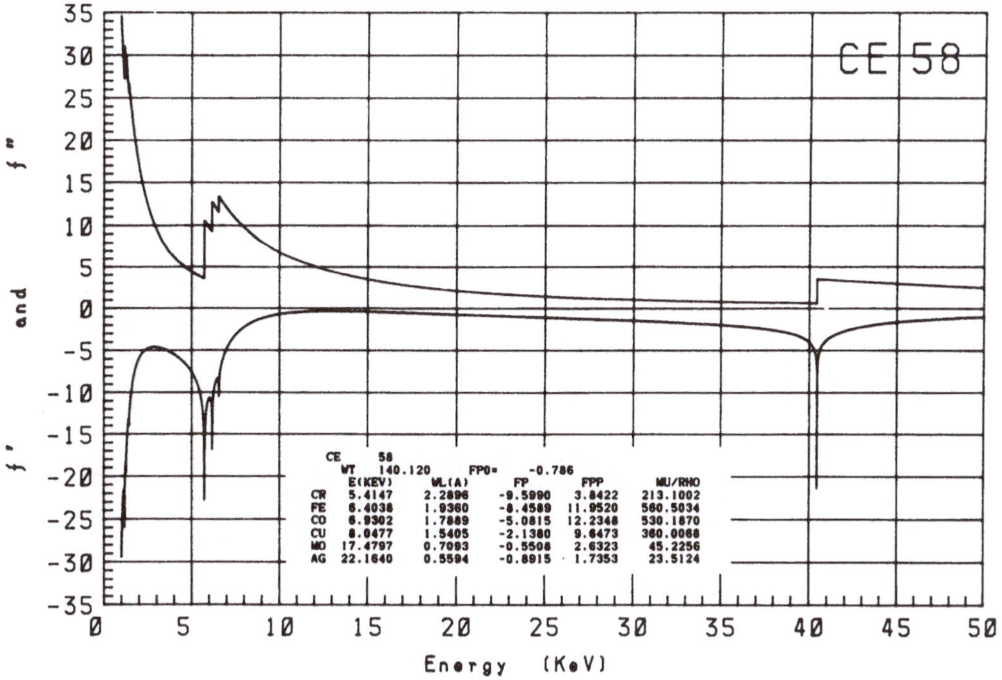
CE 58
f″ and f′
35
30
25
20
15
10
5
0
-5
-10
-15
-20
-25
-30
-35
0 5 10 15 20 25 30 35 40 45 50
Energy (KeV)
CE 58
WT 140.120 FPO= -0.786
E(KEV) WL(A) FP FPP MU/RHO
CR 5.4147 2.2896 -9.5990 3.8422 213.1002
FE 6.4038 1.9360 -8.4589 11.9520 560.5034
CO 6.9302 1.7889 -5.0815 12.2348 530.1870
CU 8.0477 1.5405 -2.1380 9.6473 360.0068
MO 17.4797 0.7093 -0.5508 2.6323 45.2256
AG 22.1640 0.5594 -0.8915 1.7353 23.5124

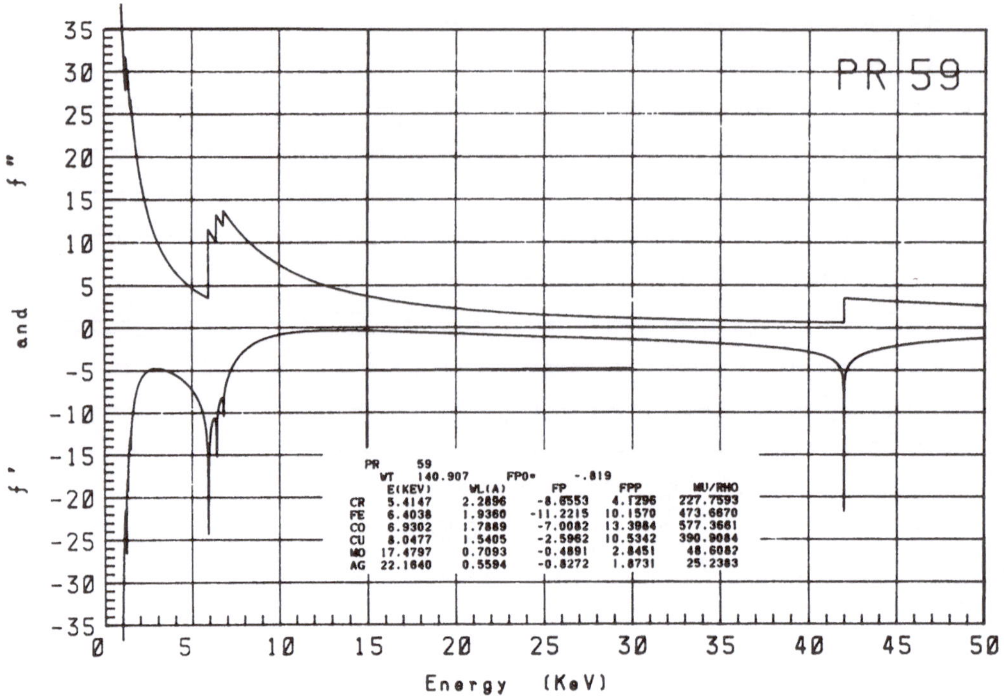

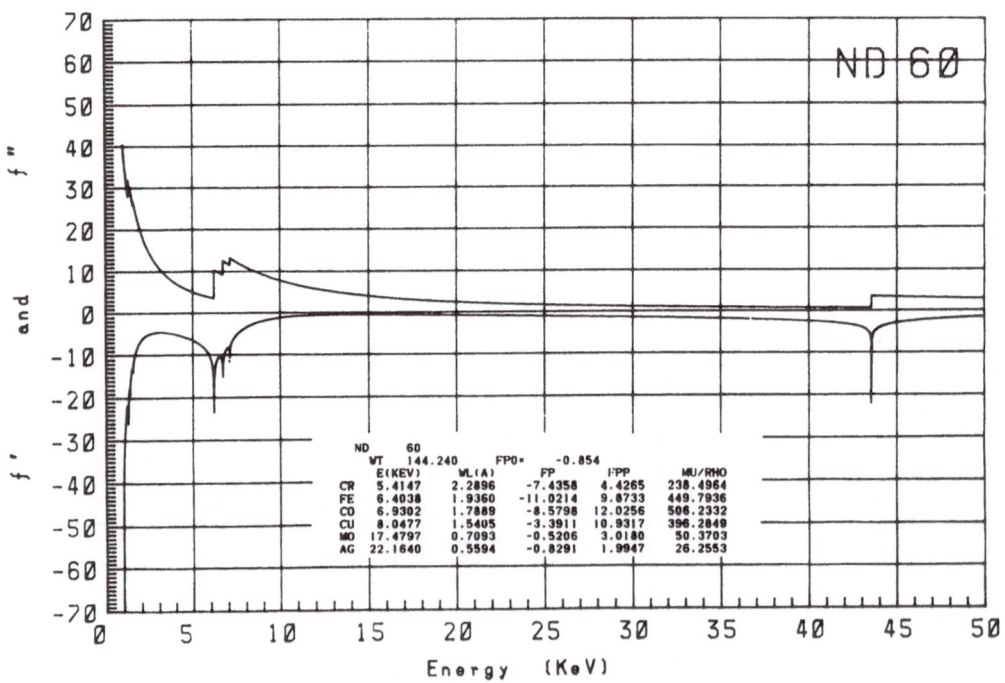

PR 59
f" and f'
Energy (KeV)
PR 59
WT 140.907 FPO= -.819

	E(KEV)	WL(A)	FP	FPP	MU/RHO
CR	5.4147	2.2896	-8.6553	4.1296	227.7593
FE	6.4038	1.9360	-11.2215	10.1570	473.6670
CO	6.9302	1.7889	-7.0082	13.3984	577.3661
CU	8.0477	1.5405	-2.5962	10.5342	390.9084
MO	17.4797	0.7093	-0.4891	2.8451	48.6082
AG	22.1640	0.5594	-0.8272	1.8731	25.2383

ND 60
f" and f'
Energy (KeV)
ND 60
WT 144.240 FPO= -0.854

	E(KEV)	WL(A)	FP	FPP	MU/RHO
CR	5.4147	2.2896	-7.4358	4.4265	238.4964
FE	6.4038	1.9360	-11.0214	9.8733	449.7936
CO	6.9302	1.7889	-8.5798	12.0256	506.2332
CU	8.0477	1.5405	-3.3911	10.9317	396.2849
MO	17.4797	0.7093	-0.5206	3.0180	50.3703
AG	22.1640	0.5594	-0.8291	1.9947	26.2553

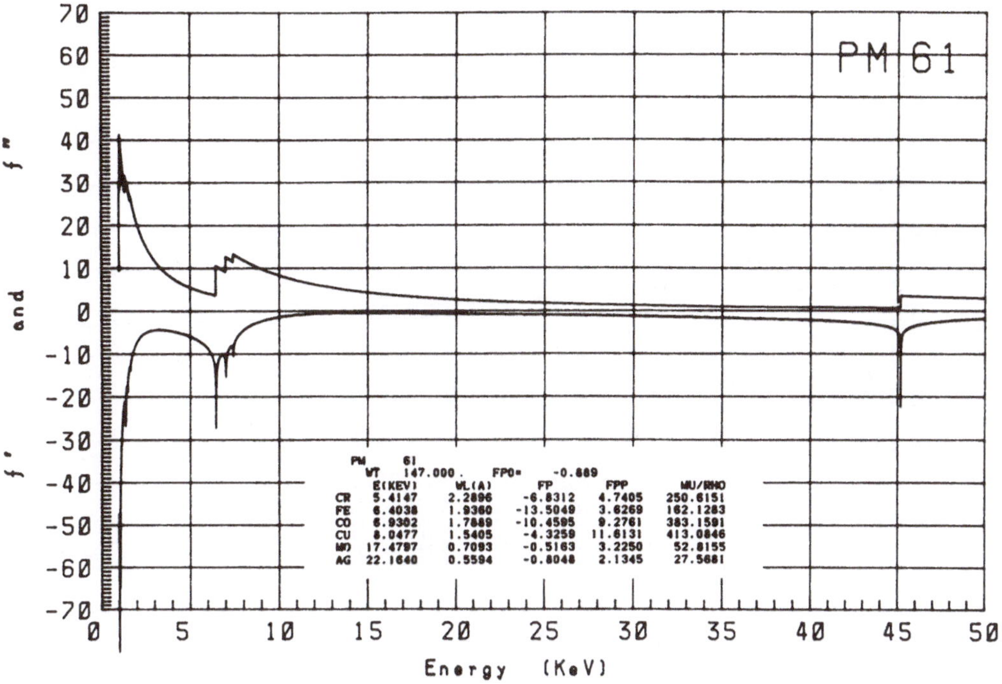
PM 61
f'' and f'
Energy (KeV)
PM 61
WT 147.000. FPO= -0.889
E(KEV) WL(A) FP FPP MU/RHO
CR 5.4147 2.2896 -6.8312 4.7405 250.6151
FE 6.4038 1.9360 -13.5049 3.6269 162.1283
CO 6.9302 1.7889 -10.4595 9.2761 383.1591
CU 8.0477 1.5405 -4.3259 11.6131 413.0846
MO 17.4797 0.7093 -0.5163 3.2250 52.8155
AG 22.1640 0.5594 -0.8048 2.1345 27.5681

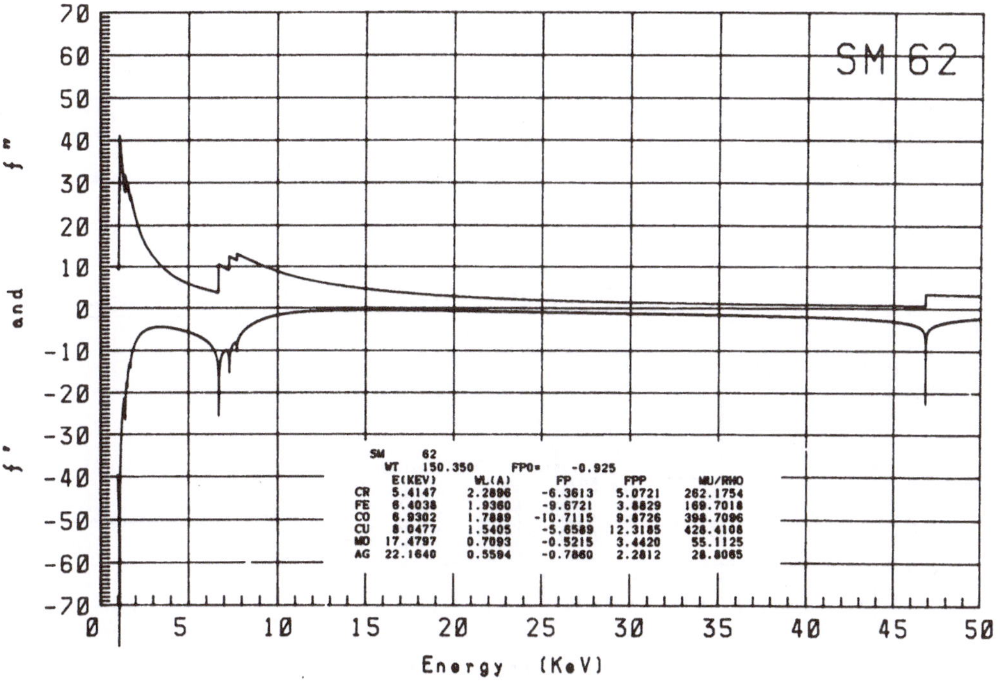
SM 62
f'' and f'
Energy (KeV)
SM 62
WT 150.350 FPO= -0.925
E(KEV) WL(A) FP FPP MU/RHO
CR 5.4147 2.2896 -6.3613 5.0721 262.1754
FE 6.4038 1.9360 -9.6721 3.8829 169.7018
CO 6.9302 1.7889 -10.7115 9.8726 398.7096
CU 8.0477 1.5405 -5.6589 12.3185 428.4108
MO 17.4797 0.7093 -0.5215 3.4420 55.1125
AG 22.1640 0.5594 -0.7860 2.2812 28.8065

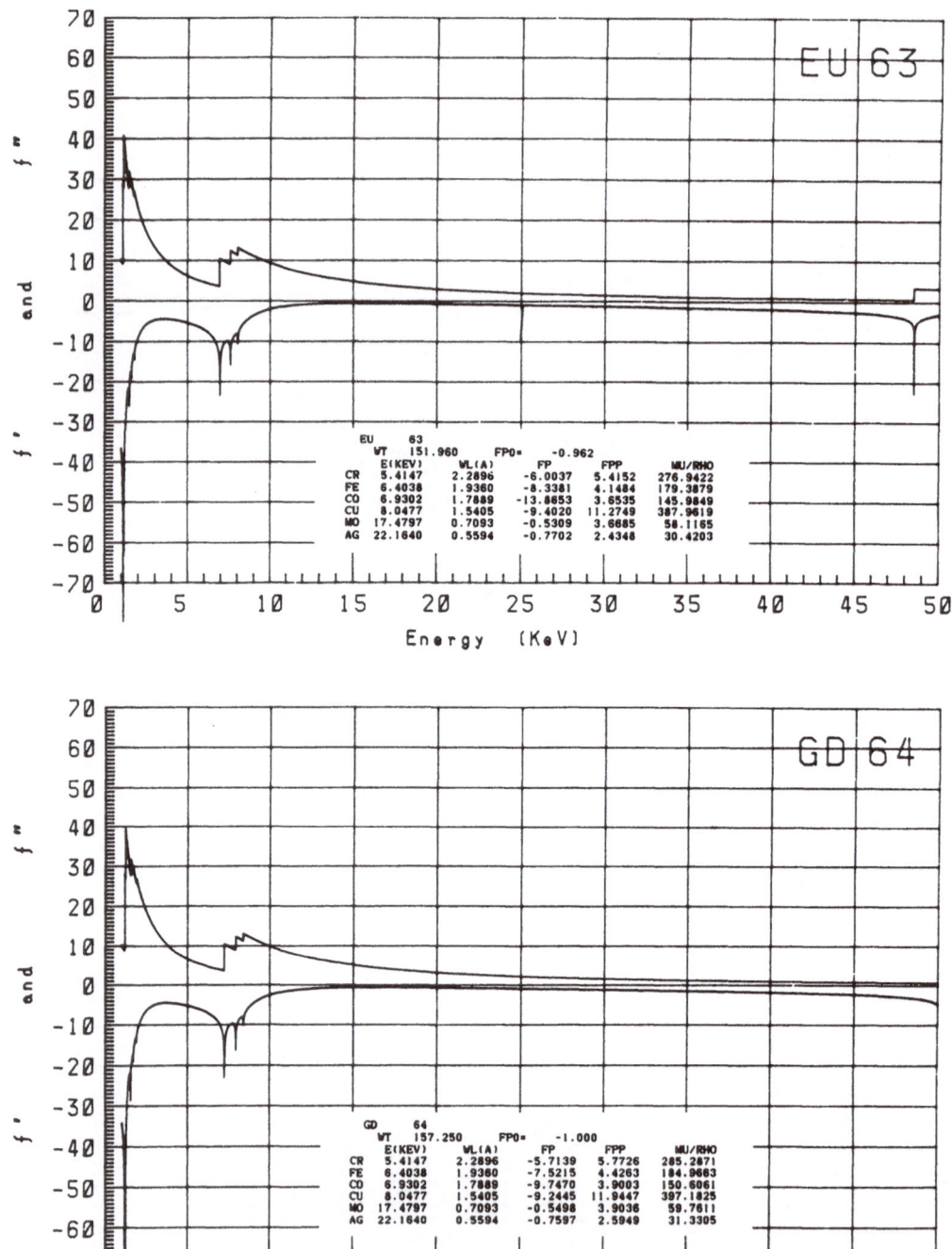

EU 63

EU	63				
WT	151.960	FP0=	-0.962		
	E(KEV)	WL(A)	FP	FPP	MU/RHO
CR	5.4147	2.2896	-6.0037	5.4152	276.9422
FE	6.4038	1.9360	-8.3381	4.1484	179.3879
CO	6.9302	1.7889	-13.8653	3.6535	145.9849
CU	8.0477	1.5405	-9.4020	11.2749	387.9619
MO	17.4797	0.7093	-0.5309	3.6685	58.1165
AG	22.1640	0.5594	-0.7702	2.4348	30.4203

GD 64

GD	64				
WT	157.250	FP0=	-1.000		
	E(KEV)	WL(A)	FP	FPP	MU/RHO
CR	5.4147	2.2896	-5.7139	5.7726	285.2871
FE	6.4038	1.9360	-7.5215	4.4263	184.9663
CO	6.9302	1.7889	-9.7470	3.9003	150.6061
CU	8.0477	1.5405	-9.2445	11.9447	397.1825
MO	17.4797	0.7093	-0.5498	3.9036	59.7611
AG	22.1640	0.5594	-0.7597	2.5949	31.3305

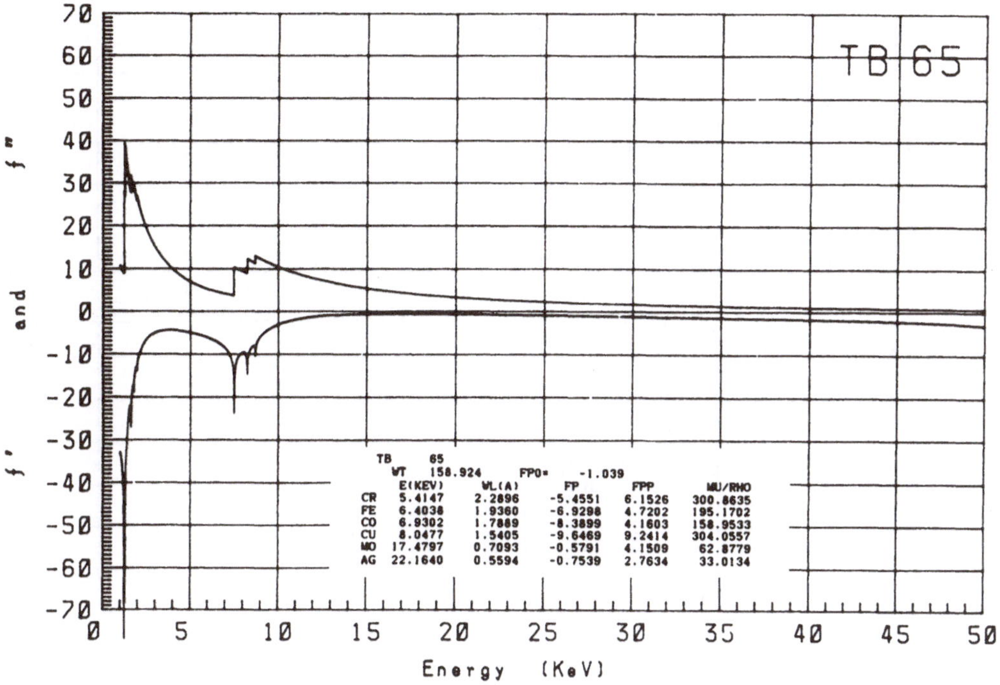
TB 65
f'' and f'
70
60
50
40
30
20
10
0
-10
-20
-30
-40
-50
-60
-70
0
5
10
15
20
25
30
35
40
45
50
Energy (KeV)
TB 65
WT 158.924 FPO= -1.039
E(KEV) WL(A) FP FPP MU/RHO
CR 5.4147 2.2896 -5.4551 6.1526 300.8635
FE 6.4038 1.9360 -6.9298 4.7202 195.1702
CO 6.9302 1.7889 -8.3899 4.1603 158.9533
CU 8.0477 1.5405 -9.6469 9.2414 304.0557
MO 17.4797 0.7093 -0.5791 4.1509 62.8779
AG 22.1640 0.5594 -0.7539 2.7634 33.0134

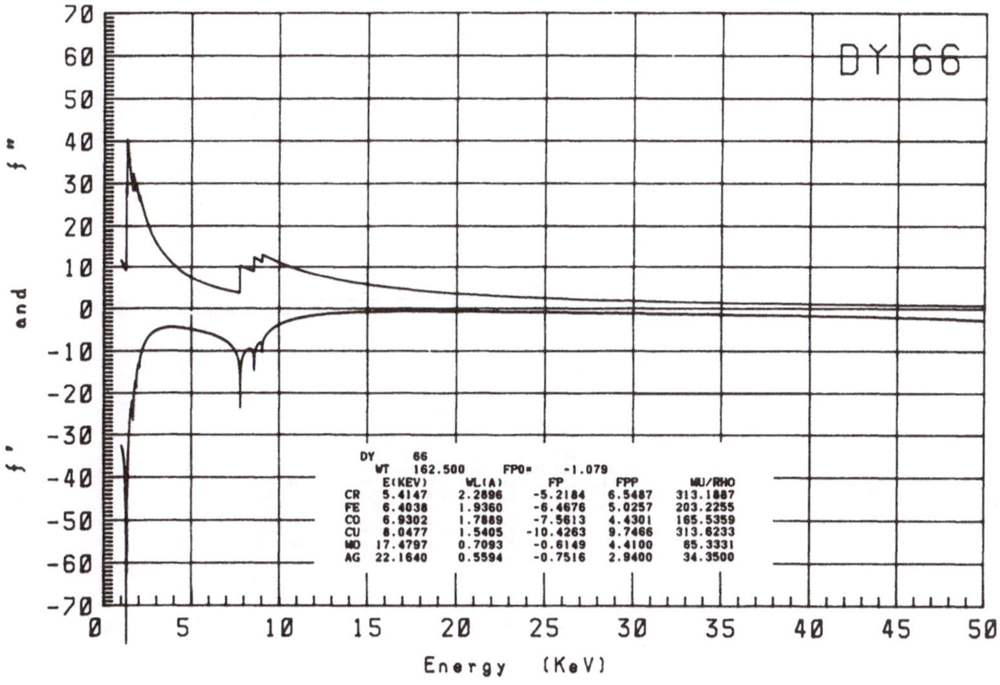
DY 66
f'' and f'
70
60
50
40
30
20
10
0
-10
-20
-30
-40
-50
-60
-70
0
5
10
15
20
25
30
35
40
45
50
Energy (KeV)
DY 66
WT 162.500 FPO= -1.079
E(KEV) WL(A) FP FPP MU/RHO
CR 5.4147 2.2896 -5.2184 6.5487 313.1887
FE 6.4038 1.9360 -6.4676 5.0257 203.2255
CO 6.9302 1.7889 -7.5613 4.4301 165.5359
CU 8.0477 1.5405 -10.4263 9.7466 313.6233
MO 17.4797 0.7093 -0.6149 4.4100 65.3331
AG 22.1640 0.5594 -0.7516 2.9400 34.3500

154

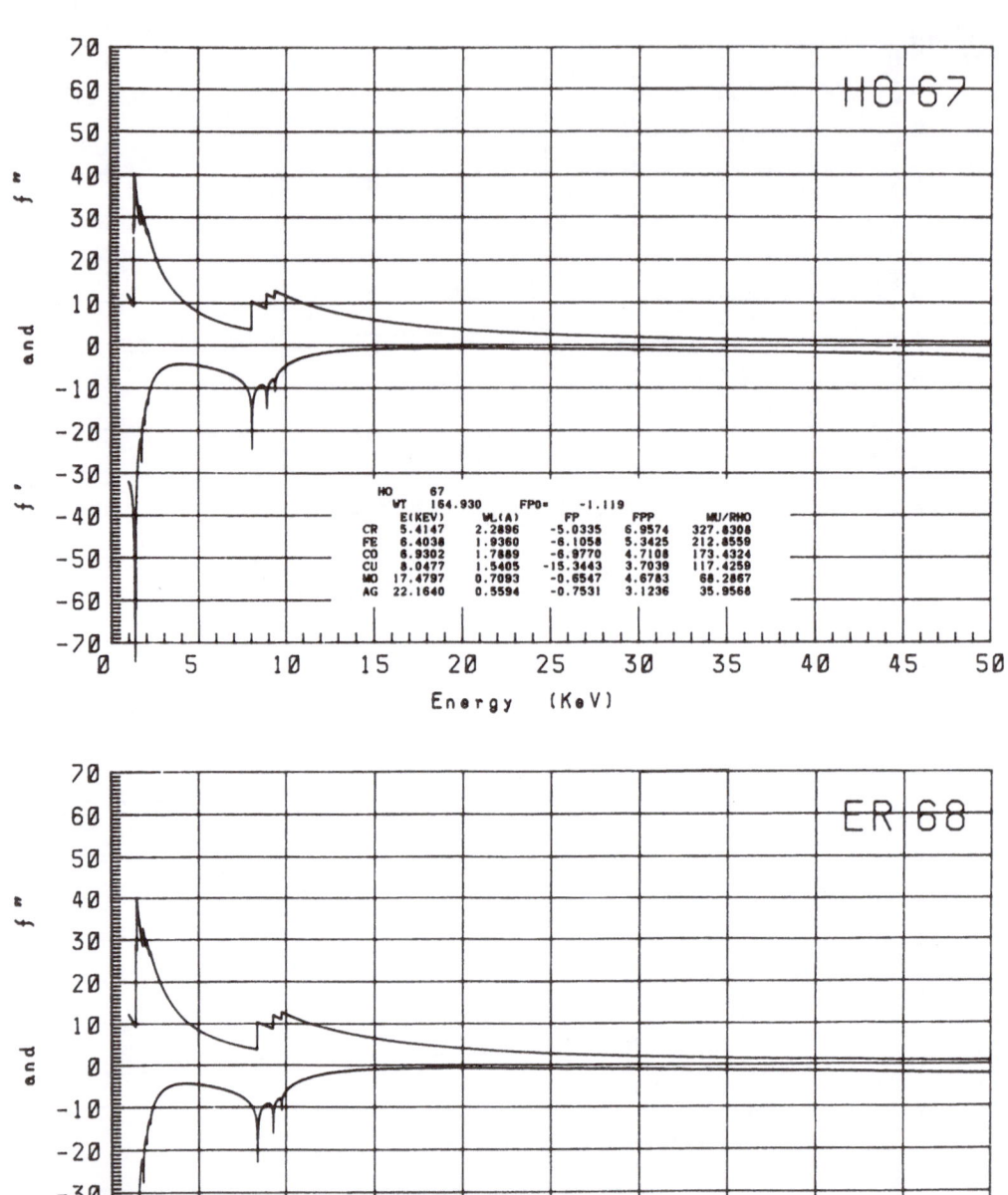

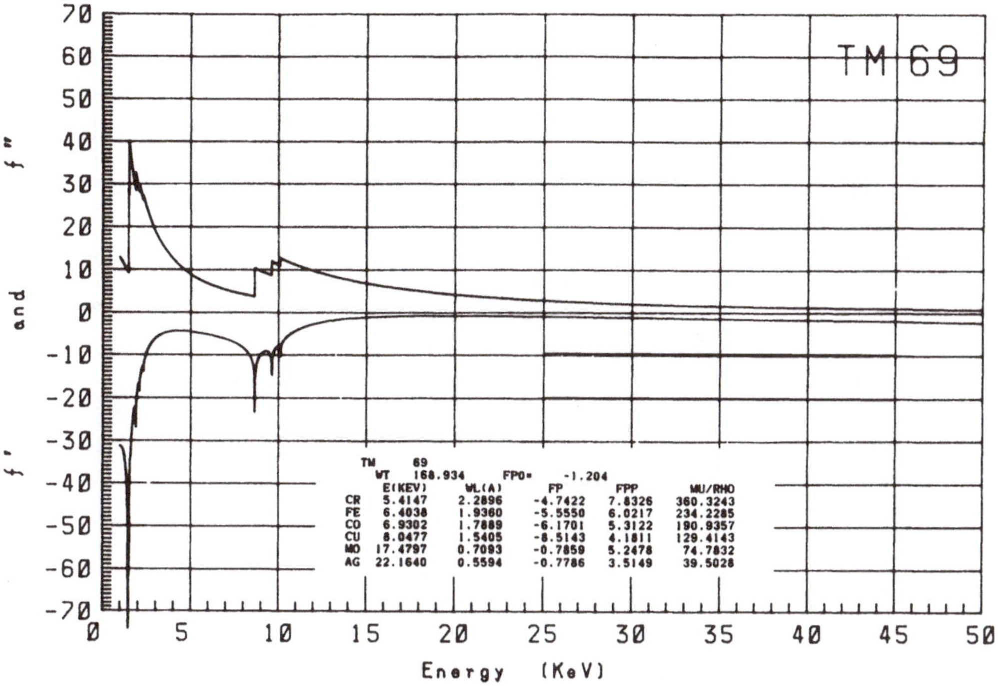

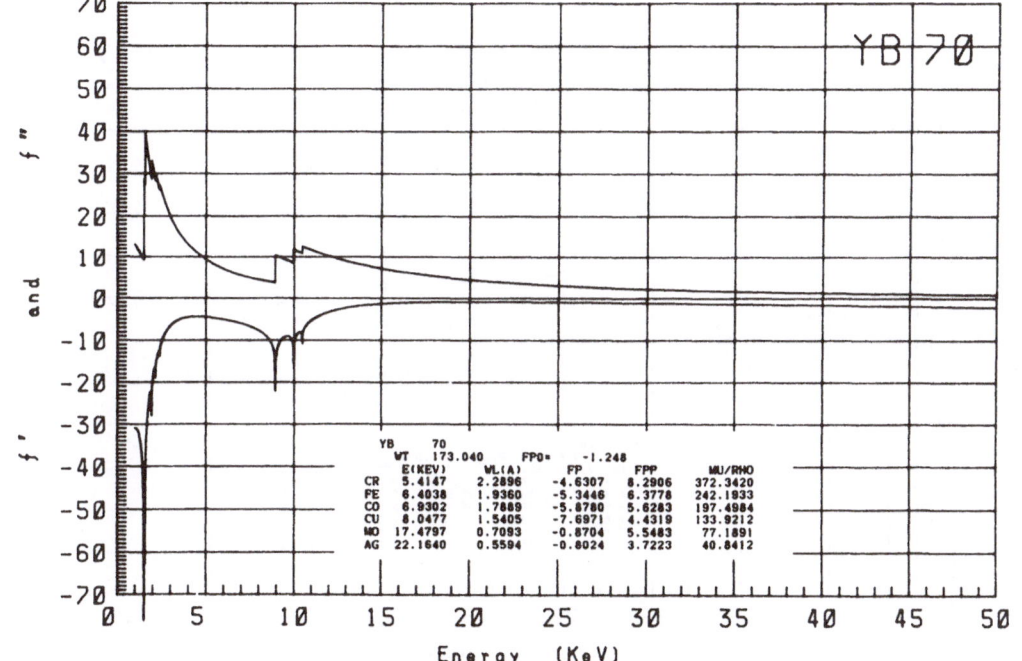

TM 69
70
60
50
40
30
20
10
0
-10
-20
-30
-40
-50
-60
-70
f" and f'
TM 69
WT 168.934 FPO= -1.204
E(KEV) WL(A) FP FPP MU/RHO
CR 5.4147 2.2896 -4.7422 7.8326 360.3243
FE 6.4038 1.9360 -5.5550 6.0217 234.2285
CO 6.9302 1.7889 -6.1701 5.3122 190.9357
CU 8.0477 1.5405 -8.5143 4.1811 129.4143
MO 17.4797 0.7093 -0.7859 5.2478 74.7832
AG 22.1640 0.5594 -0.7786 3.5149 39.5028
0
5
10
15
20
25
30
35
40
45
50
Energy (KeV)

YB 70
70
60
50
40
30
20
10
0
-10
-20
-30
-40
-50
-60
-70
f" and f'
YB 70
WT 173.040 FPO= -1.248
E(KEV) WL(A) FP FPP MU/RHO
CR 5.4147 2.2896 -4.6307 8.2906 372.3420
FE 6.4038 1.9360 -5.3446 6.3778 242.1933
CO 6.9302 1.7889 -5.8780 5.6283 197.4984
CU 8.0477 1.5405 -7.6971 4.4319 133.9212
MO 17.4797 0.7093 -0.8704 5.5483 77.1891
AG 22.1640 0.5594 -0.8024 3.7223 40.8412
0
5
10
15
20
25
30
35
40
45
50
Energy (KeV)

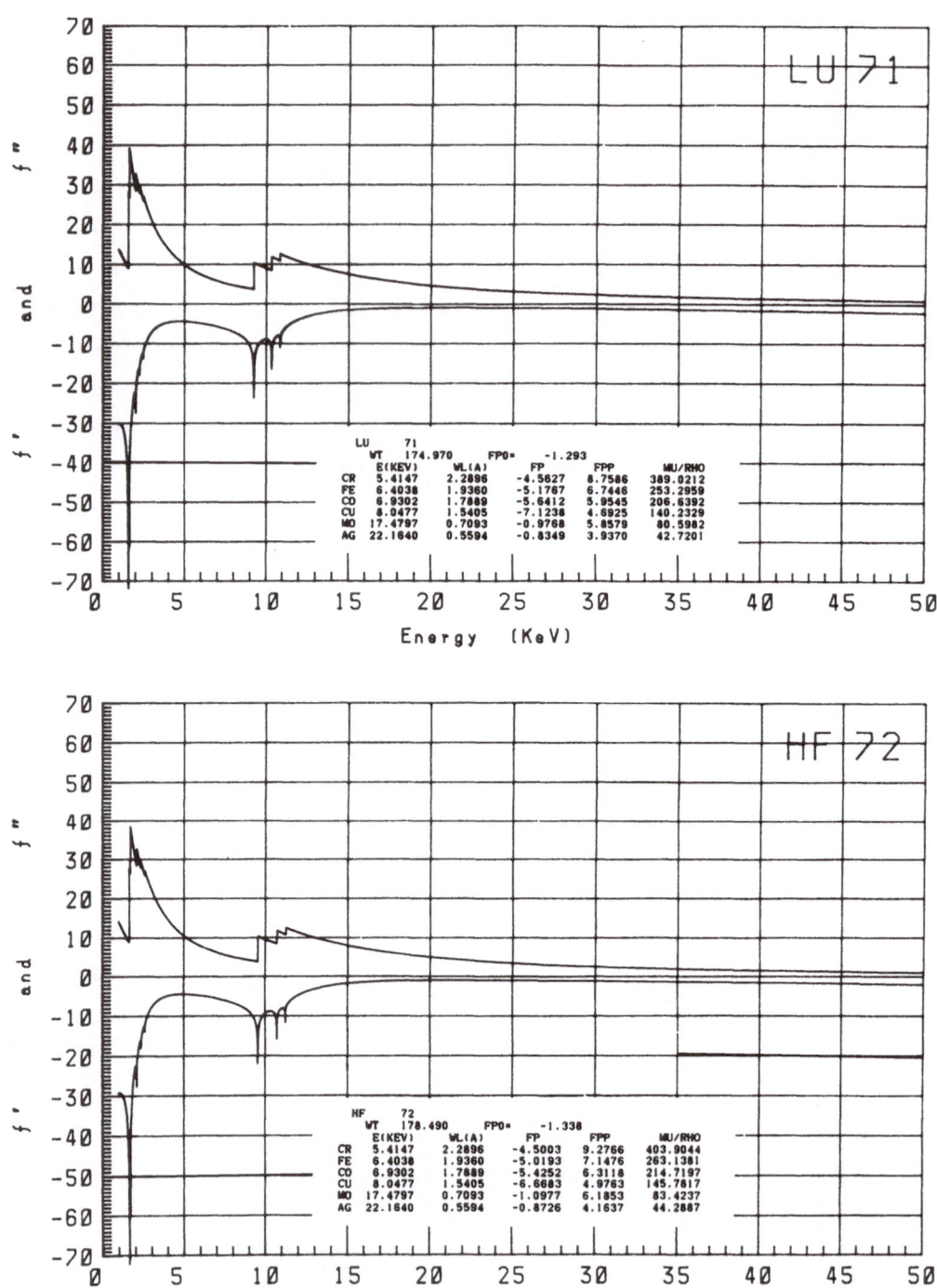

LU 71

LU	71				
WT	174.970		FP0=	-1.293	
	E(KEV)	WL(A)	FP	FPP	MU/RHO
CR	5.4147	2.2896	-4.5627	8.7586	389.0212
FE	6.4038	1.9360	-5.1767	6.7446	253.2959
CO	6.9302	1.7889	-5.6412	5.9545	206.6392
CU	8.0477	1.5405	-7.1238	4.6925	140.2329
MO	17.4797	0.7093	-0.9768	5.8579	80.5982
AG	22.1640	0.5594	-0.8349	3.9370	42.7201

HF 72

HF	72				
WT	178.490		FP0=	-1.338	
	E(KEV)	WL(A)	FP	FPP	MU/RHO
CR	5.4147	2.2896	-4.5003	9.2766	403.9044
FE	6.4038	1.9360	-5.0193	7.1476	263.1381
CO	6.9302	1.7889	-5.4252	6.3118	214.7197
CU	8.0477	1.5405	-6.6683	4.9763	145.7617
MO	17.4797	0.7093	-1.0977	6.1853	83.4237
AG	22.1640	0.5594	-0.8726	4.1637	44.2887

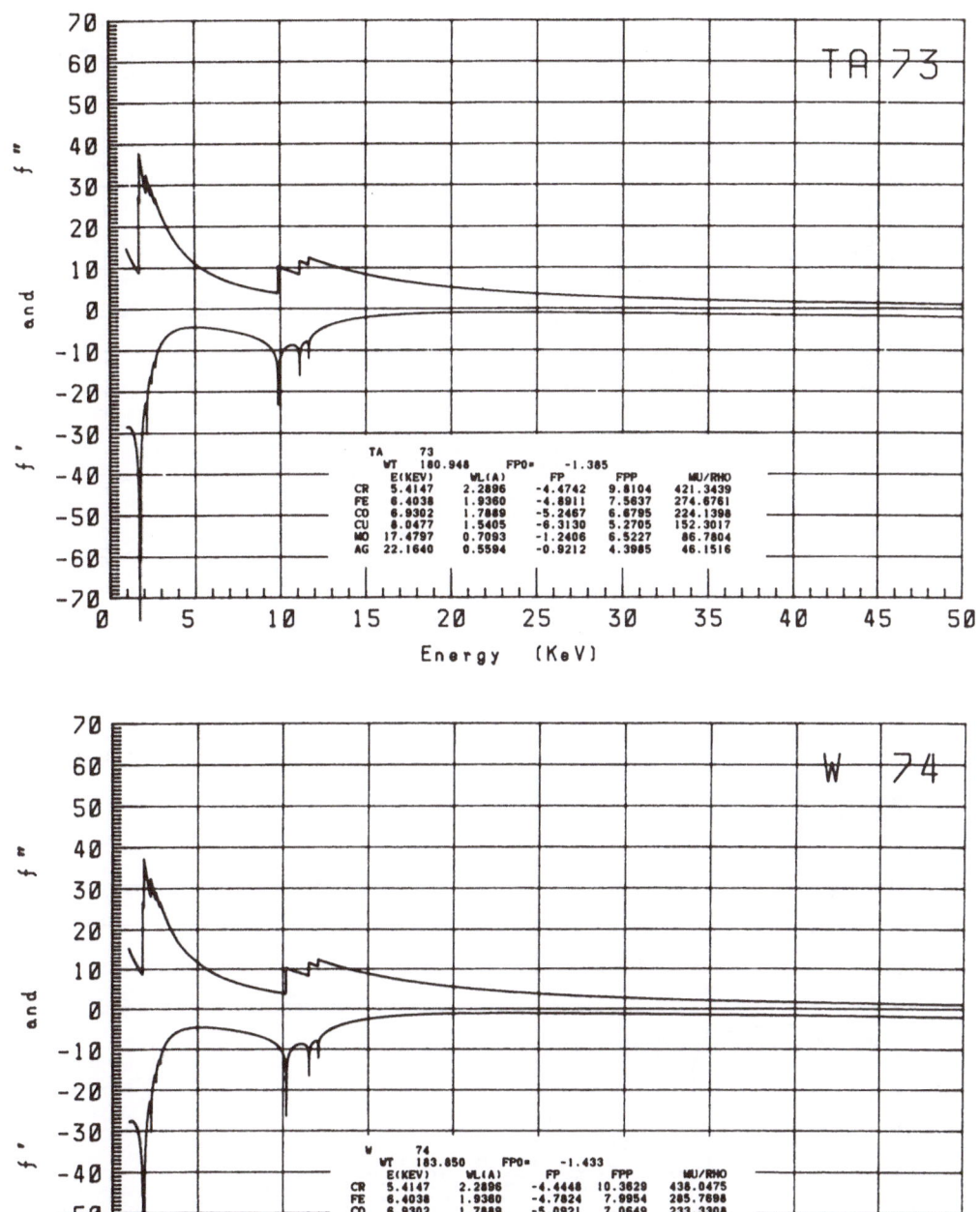

TA 73

	TA	73	FPO=	-1.385	
	WT	180.948			
	E(KEV)	WL(A)	FP	FPP	MU/RHO
CR	5.4147	2.2896	-4.4742	9.8104	421.3439
FE	6.4038	1.9360	-4.8911	7.5637	274.6761
CO	6.9302	1.7889	-5.2467	6.6795	224.1398
CU	8.0477	1.5405	-6.3130	5.2705	152.3017
MO	17.4797	0.7093	-1.2406	6.5227	86.7804
AG	22.1640	0.5594	-0.9212	4.3985	46.1516

W 74

	W	74	FPO=	-1.433	
	WT	183.850			
	E(KEV)	WL(A)	FP	FPP	MU/RHO
CR	5.4147	2.2896	-4.4448	10.3629	438.0475
FE	6.4038	1.9360	-4.7824	7.9954	285.7898
CO	6.9302	1.7889	-5.0921	7.0649	233.3308
CU	8.0477	1.5405	-6.0179	5.5761	158.5900
MO	17.4797	0.7093	-1.4054	6.8721	89.9852
AG	22.1640	0.5594	-0.9786	4.6423	47.9408

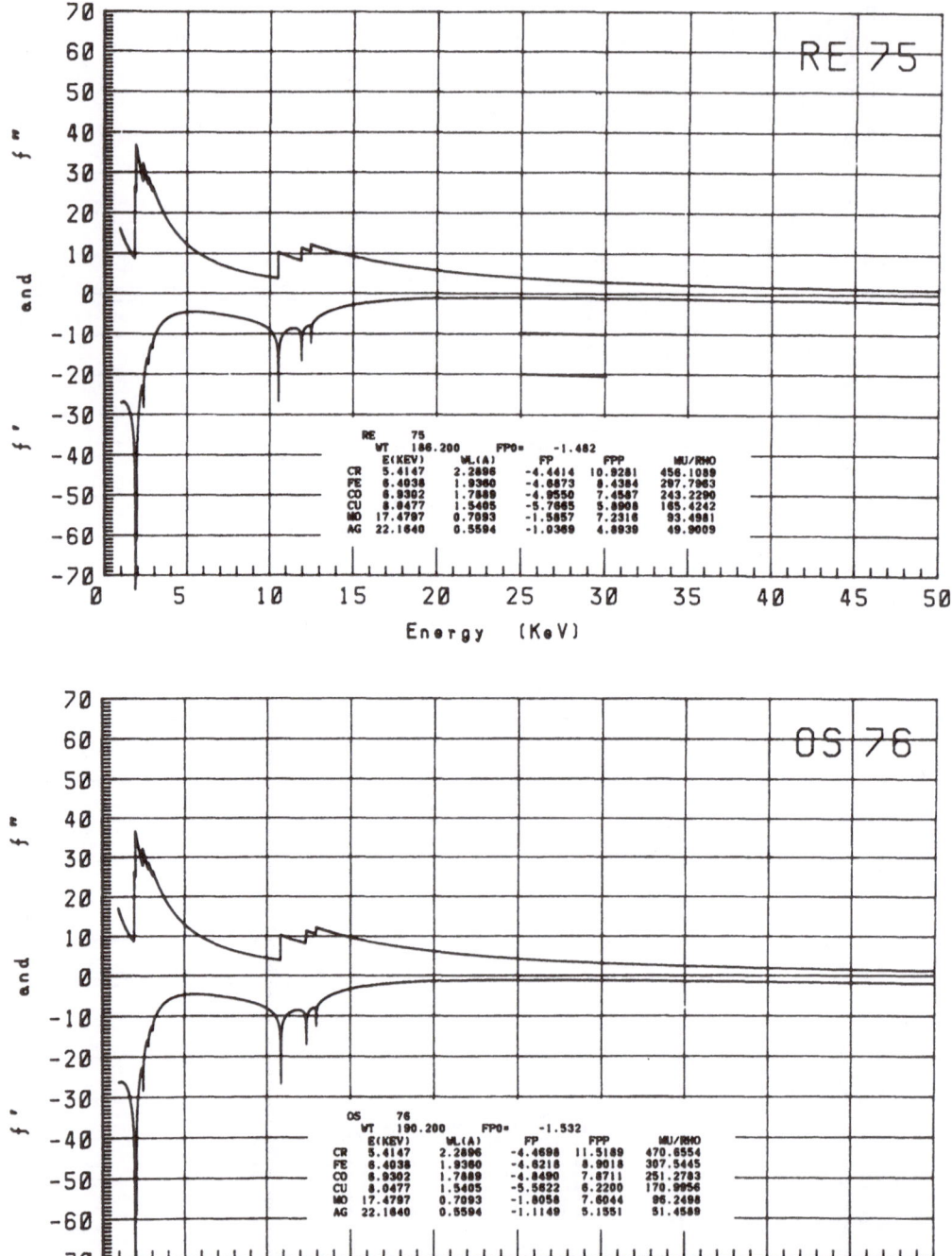

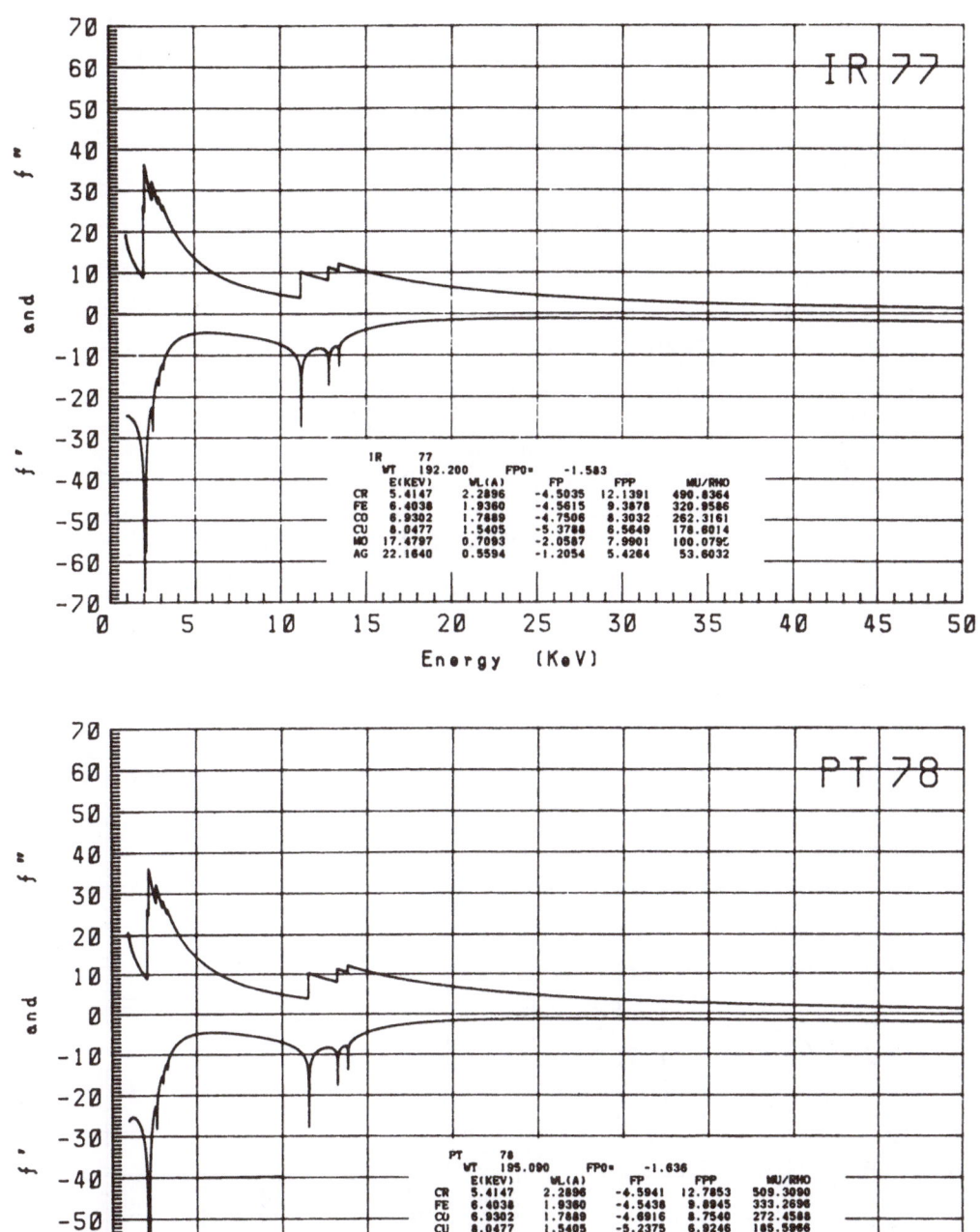

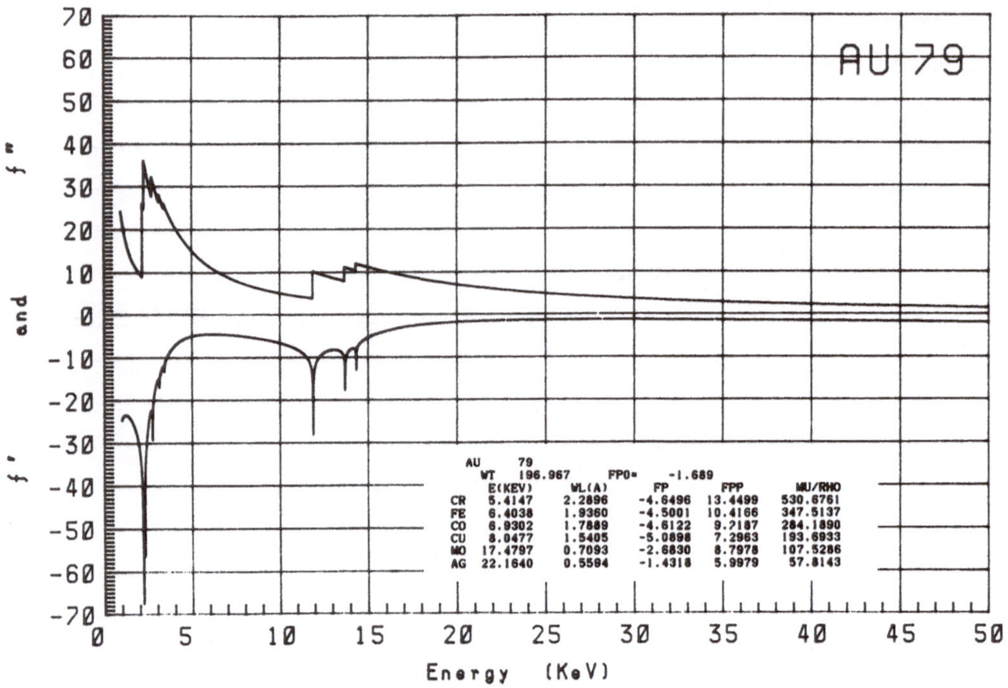

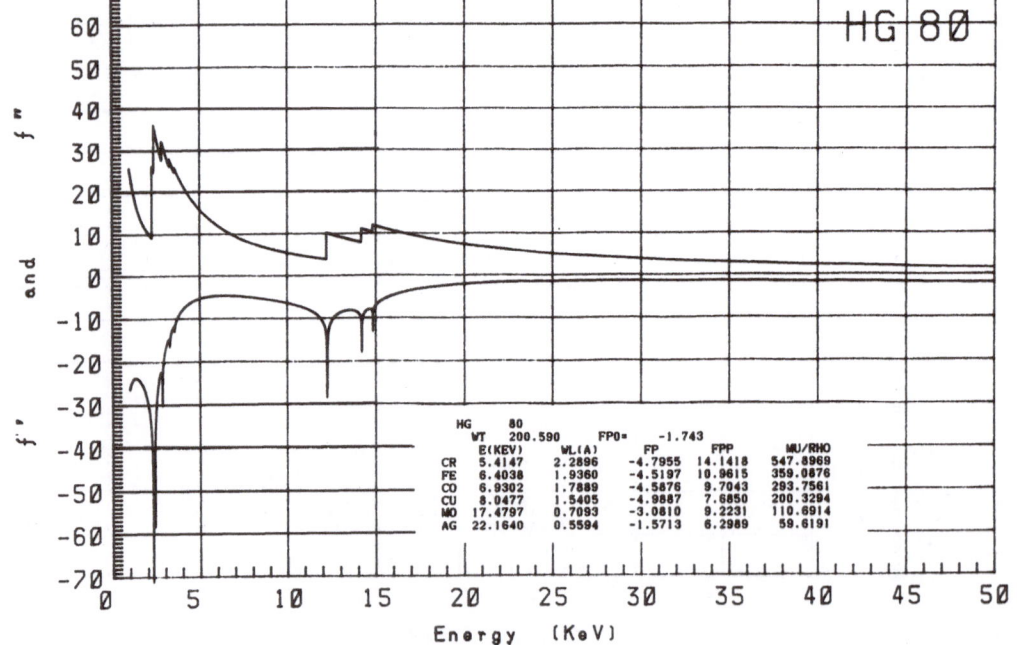

AU   79
WT   196.967   FPO=   -1.689
	E(KEV)	WL(A)	FP	FPP	MU/RHO
CR	5.4147	2.2896	-4.6496	13.4499	530.6761
FE	6.4038	1.9360	-4.5001	10.4166	347.5137
CO	6.9302	1.7889	-4.6122	9.2187	284.1890
CU	8.0477	1.5405	-5.0898	7.2963	193.6933
MO	17.4797	0.7093	-2.6830	8.7978	107.5286
AG	22.1640	0.5594	-1.4318	5.9979	57.8143

HG   80
WT   200.590   FPO=   -1.743
	E(KEV)	WL(A)	FP	FPP	MU/RHO
CR	5.4147	2.2896	-4.7955	14.1418	547.8969
FE	6.4038	1.9360	-4.5197	10.9615	359.0876
CO	6.9302	1.7889	-4.5876	9.7043	293.7561
CU	8.0477	1.5405	-4.9887	7.6850	200.3294
MO	17.4797	0.7093	-3.0810	9.2231	110.6914
AG	22.1640	0.5594	-1.5713	6.2989	59.6191

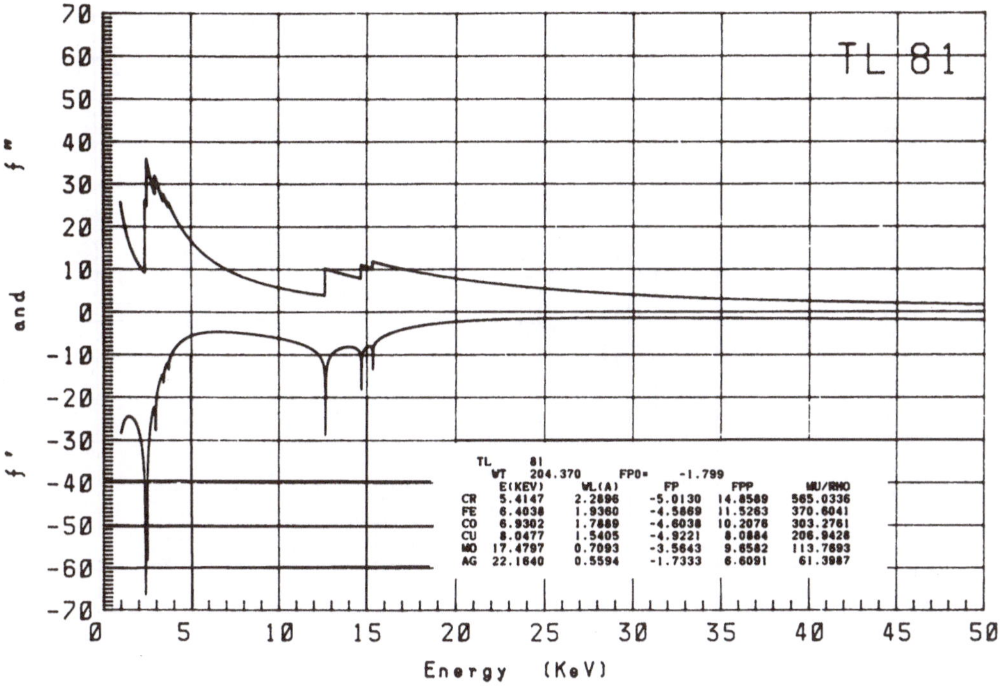

TL 81

	E(KEV)	WL(A)	FP	FPP	MU/RHO
TL 81					
	WT 204.370		FP0= -1.799		
CR	5.4147	2.2896	-5.0130	14.8589	565.0336
FE	6.4038	1.9360	-4.5869	11.5263	370.6041
CO	6.9302	1.7889	-4.6038	10.2076	303.2761
CU	8.0477	1.5405	-4.9221	8.0884	206.9428
MO	17.4797	0.7093	-3.5643	9.8582	113.7693
AG	22.1640	0.5594	-1.7333	6.6091	61.3987

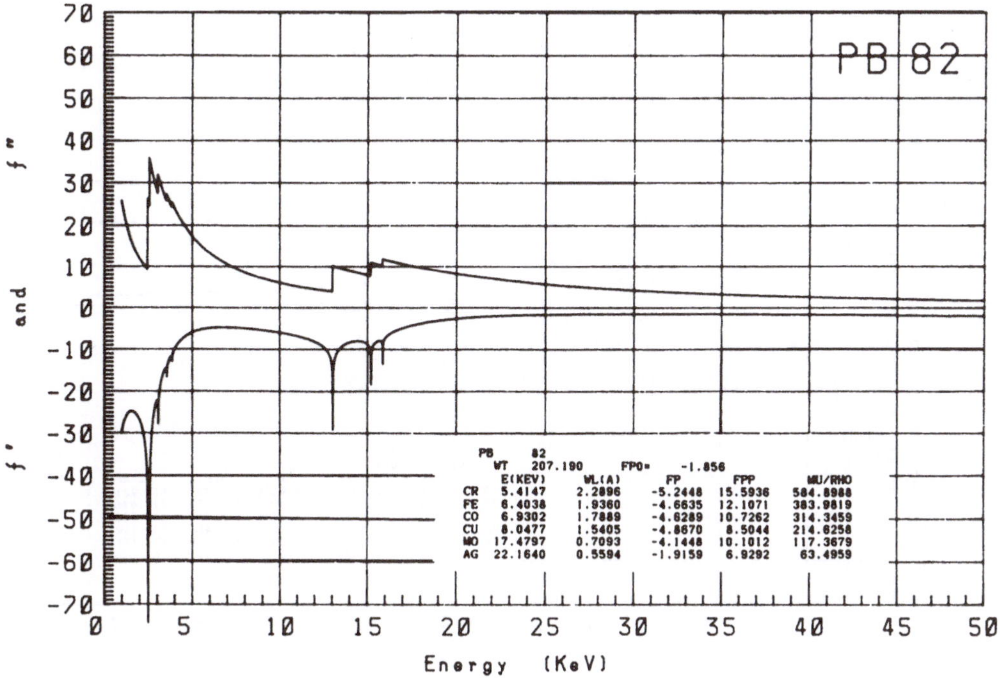

PB 82

	E(KEV)	WL(A)	FP	FPP	MU/RHO
PB 82					
	WT 207.190		FP0= -1.856		
CR	5.4147	2.2896	-5.2448	15.5936	584.8988
FE	6.4038	1.9360	-4.6635	12.1071	383.9819
CO	6.9302	1.7889	-4.8289	10.7262	314.3459
CU	8.0477	1.5405	-4.8670	8.5044	214.6258
MO	17.4797	0.7093	-4.1448	10.1012	117.3679
AG	22.1640	0.5594	-1.9159	6.9292	63.4959

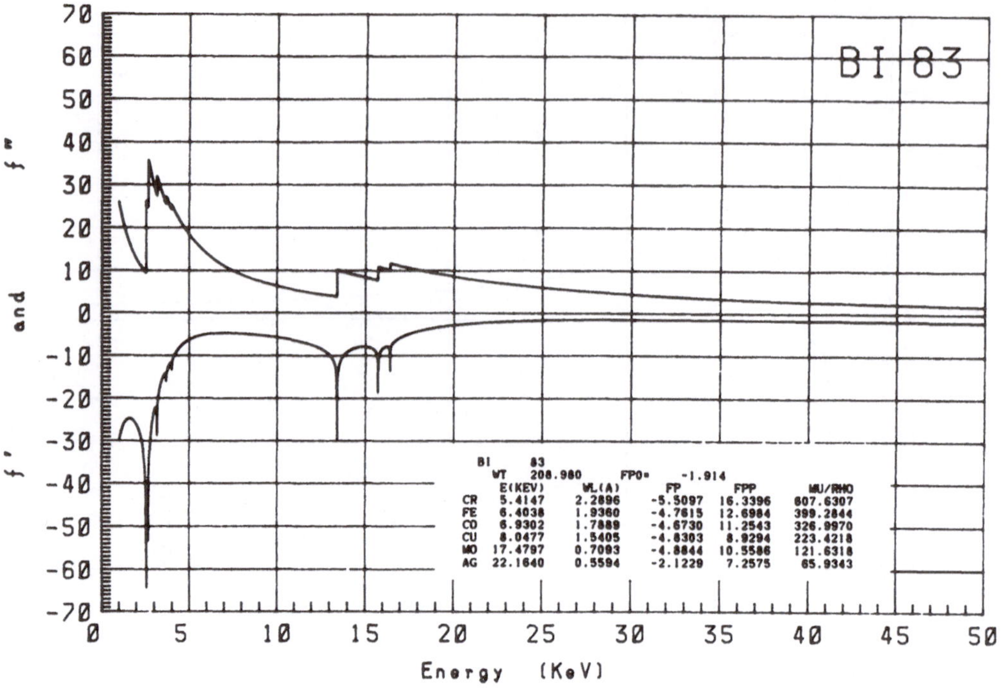
BI 83
f'' and f'
Energy (KeV)
70
60
50
40
30
20
10
0
-10
-20
-30
-40
-50
-60
-70
0
5
10
15
20
25
30
35
40
45
50
BI 83
WT 208.980 FPO= -1.914
E(KEV) WL(A) FP FPP MU/RHO
CR 5.4147 2.2896 -5.5097 16.3396 607.6307
FE 6.4038 1.9360 -4.7615 12.6984 399.2844
CO 6.9302 1.7889 -4.6730 11.2543 326.9970
CU 8.0477 1.5405 -4.8303 8.9294 223.4218
MO 17.4797 0.7093 -4.8844 10.5586 121.6318
AG 22.1640 0.5594 -2.1229 7.2575 65.9343

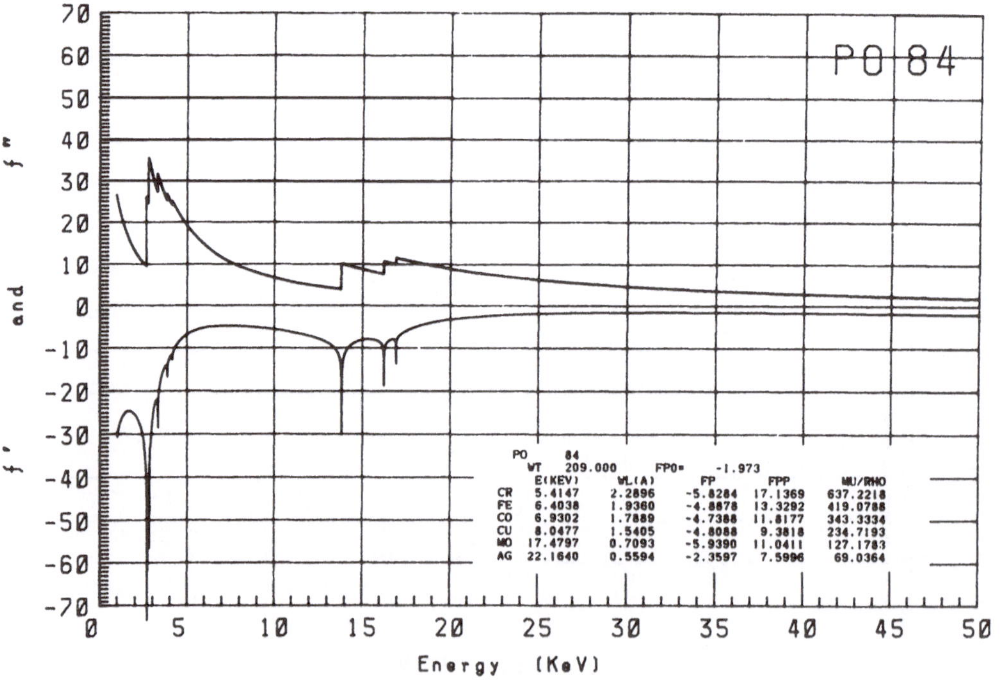
PO 84
f'' and f'
Energy (KeV)
70
60
50
40
30
20
10
0
-10
-20
-30
-40
-50
-60
-70
0
5
10
15
20
25
30
35
40
45
50
PO 84
WT 209.000 FPO= -1.973
E(KEV) WL(A) FP FPP MU/RHO
CR 5.4147 2.2896 -5.8284 17.1369 637.2218
FE 6.4038 1.9360 -4.8878 13.3292 419.0788
CO 6.9302 1.7889 -4.7388 11.8177 343.3334
CU 8.0477 1.5405 -4.8088 9.3818 234.7193
MO 17.4797 0.7093 -5.9390 11.0411 127.1783
AG 22.1640 0.5594 -2.3597 7.5996 69.0364

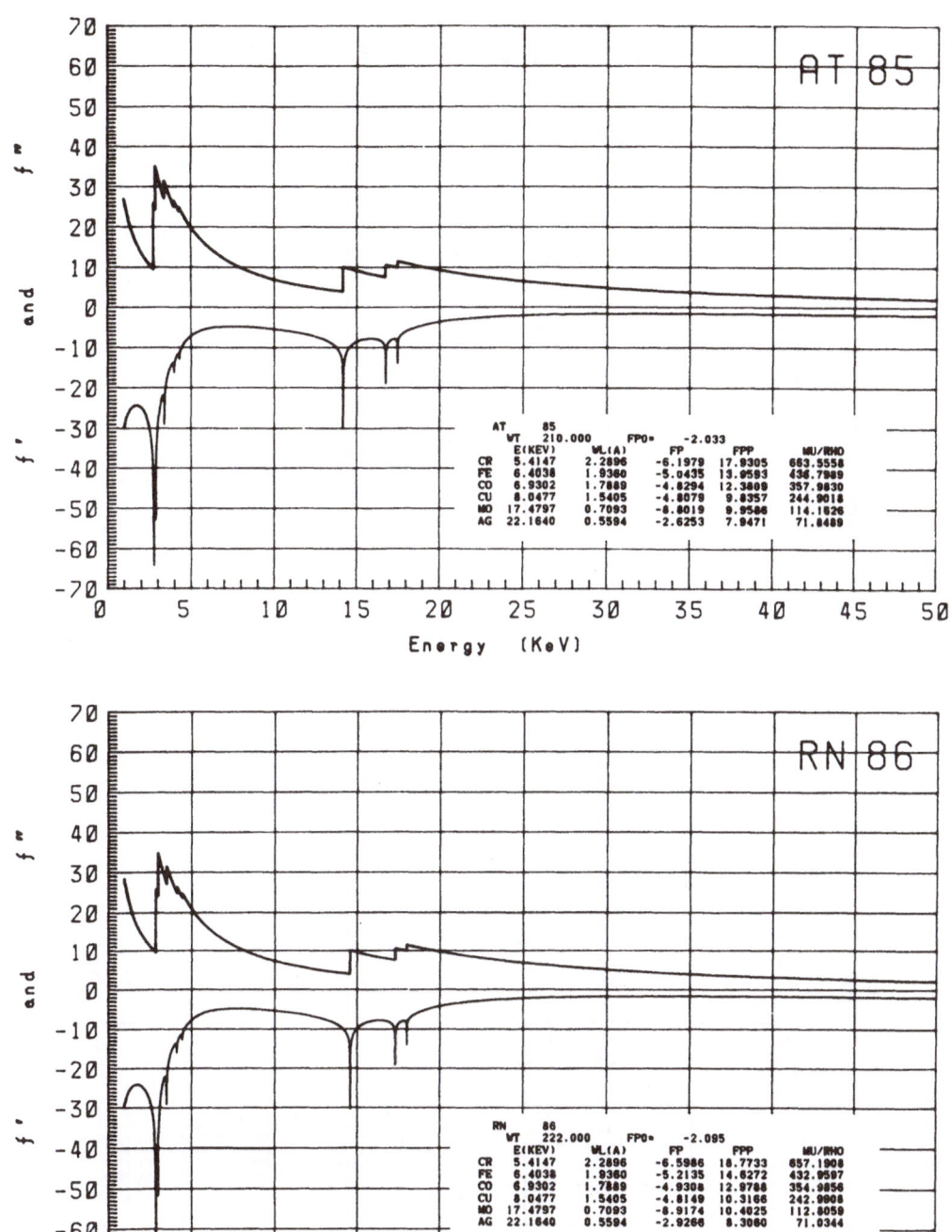

AT 85

AT	85				
WT	210.000		FP0=	-2.033	
	E(KEV)	WL(A)	FP	FPP	MU/RHO
CR	5.4147	2.2896	-6.1979	17.9305	663.5558
FE	6.4038	1.9360	-5.0435	13.9593	436.7989
CO	6.9302	1.7889	-4.8294	12.3809	357.9830
CU	8.0477	1.5405	-4.8079	9.8357	244.9018
MO	17.4797	0.7093	-8.8019	9.9586	114.1826
AG	22.1640	0.5594	-2.6253	7.9471	71.8489

RN 86

RN	86				
WT	222.000		FP0=	-2.095	
	E(KEV)	WL(A)	FP	PPP	MU/RHO
CR	5.4147	2.2896	-6.5986	18.7733	657.1908
FE	6.4038	1.9360	-5.2135	14.6272	432.9597
CO	6.9302	1.7889	-4.9308	12.9788	354.9856
CU	8.0477	1.5405	-4.8149	10.3166	242.9908
MO	17.4797	0.7093	-8.9174	10.4025	112.8059
AG	22.1640	0.5594	-2.9266	8.3080	71.0344

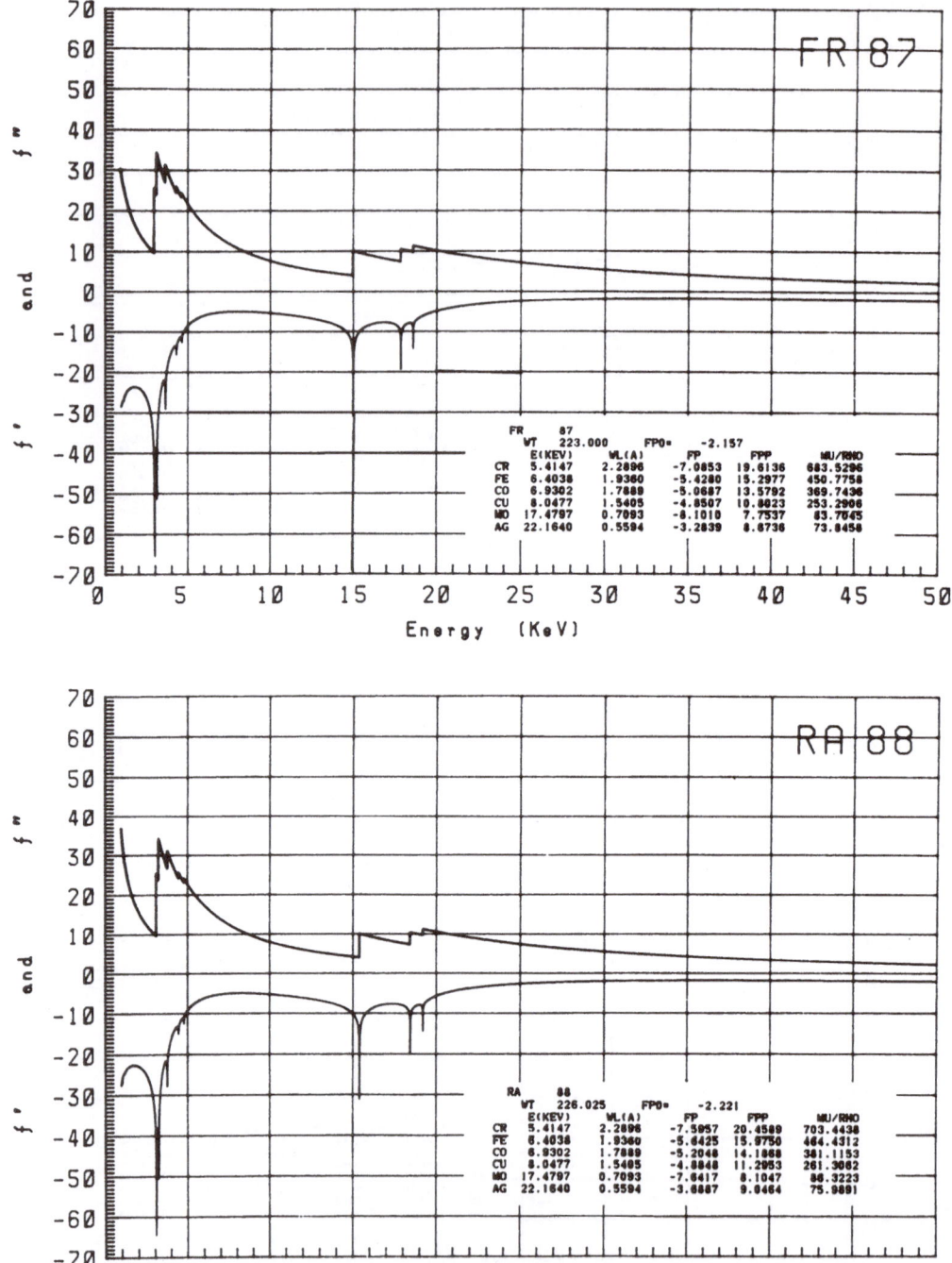

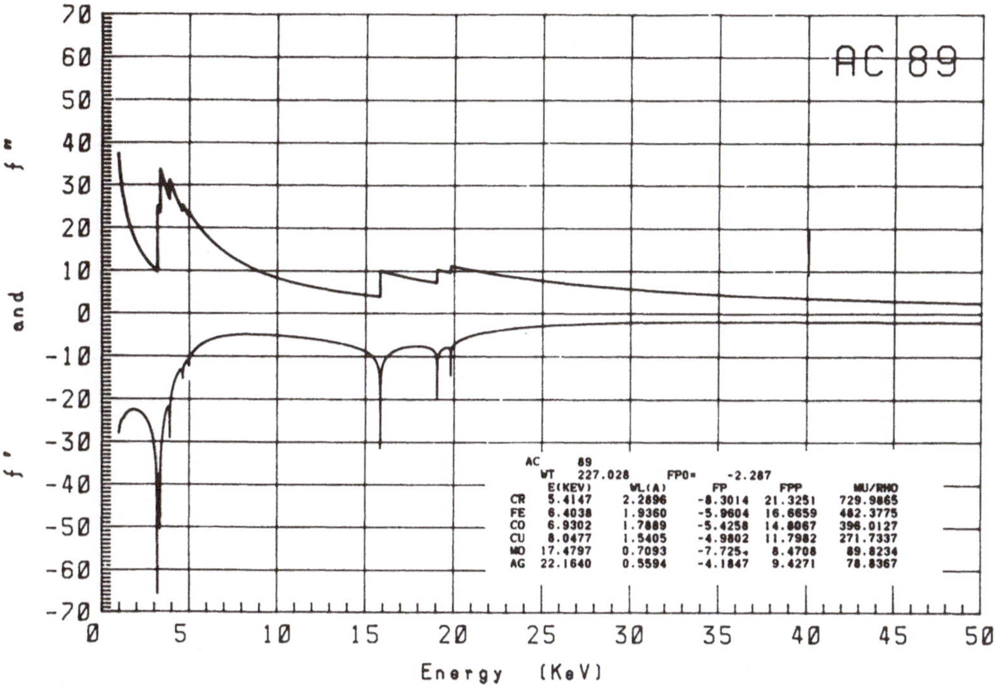
AC 89
f'' and f'
Energy (KeV)
70
60
50
40
30
20
10
0
-10
-20
-30
-40
-50
-60
-70
0
5
10
15
20
25
30
35
40
45
50
AC 89
WT 227.028 FPO= -2.287
E(KEV) WL(A) FP FPP MU/RHO
CR 5.4147 2.2896 -8.3014 21.3251 729.9865
FE 6.4038 1.9360 -5.9604 16.6659 482.3775
CO 6.9302 1.7889 -5.4258 14.8067 396.0127
CU 8.0477 1.5405 -4.9802 11.7982 271.7337
MO 17.4797 0.7093 -7.725₊ 8.4708 89.8234
AG 22.1640 0.5594 -4.1847 9.4271 78.8367

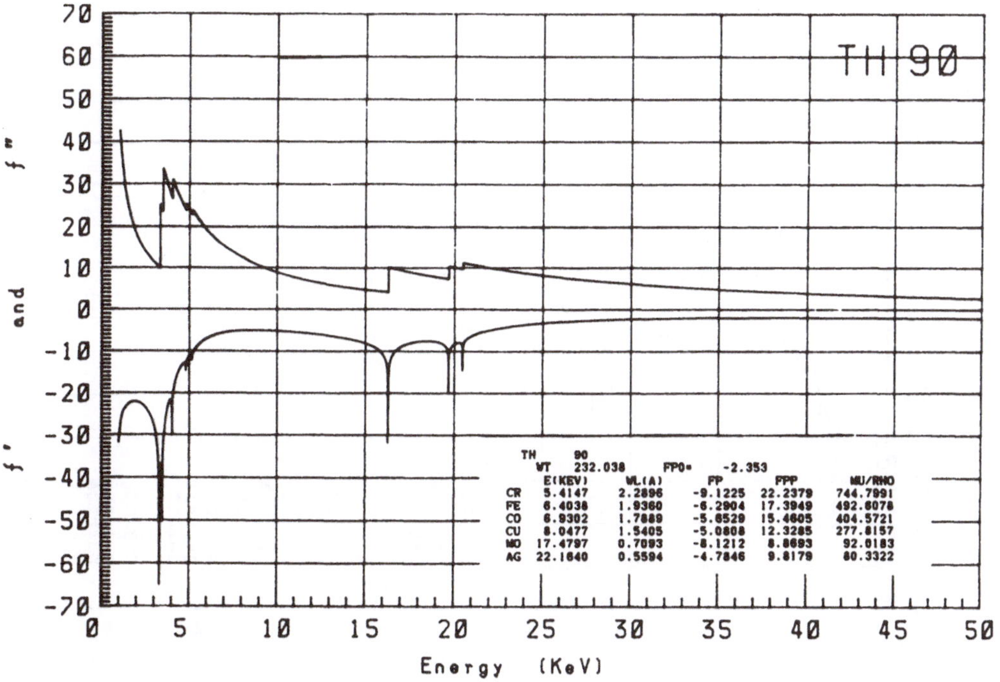
TH 90
f'' and f'
Energy (KeV)
70
60
50
40
30
20
10
0
-10
-20
-30
-40
-50
-60
-70
0
5
10
15
20
25
30
35
40
45
50
TH 90
WT 232.038 FPO= -2.353
E(KEV) WL(A) FP FPP MU/RHO
CR 5.4147 2.2896 -9.1225 22.2379 744.7991
FE 6.4038 1.9360 -6.2904 17.3949 492.6078
CO 6.9302 1.7889 -5.6529 15.4605 404.5721
CU 8.0477 1.5405 -5.0808 12.3285 277.8157
MO 17.4797 0.7093 -8.1212 8.8693 92.0183
AG 22.1640 0.5594 -4.7846 9.8179 80.3322

PA 91
WT 231.036 FPO= -2.421

	E(KEV)	WL(A)	FP	FPP	MU/RHO
CR	5.4147	2.2896	-10.3196	23.1584	778.9917
FE	6.4038	1.9360	-6.6380	18.1379	515.8767
CO	6.9302	1.7889	-5.8873	16.1262	423.8212
CU	8.0477	1.5405	-5.1753	12.8671	291.2116
MO	17.4797	0.7093	-8.9371	9.2837	96.7363
AG	22.1640	0.5594	-5.5335	10.2263	84.0372

U 92
WT 238.030 FPO= -2.490

	E(KEV)	WL(A)	FP	FPP	MU/RHO
CR	5.4147	2.2896	-10.9085	23.1191	754.8178
FE	6.4038	1.9360	-7.0618	18.8772	521.1294
CO	6.9302	1.7889	-6.1770	16.7912	428.3329
CU	8.0477	1.5405	-5.3122	13.4075	294.5261
MO	17.4797	0.7093	-10.6426	9.6539	97.6377
AG	22.1640	0.5594	-6.7036	10.6363	84.8383

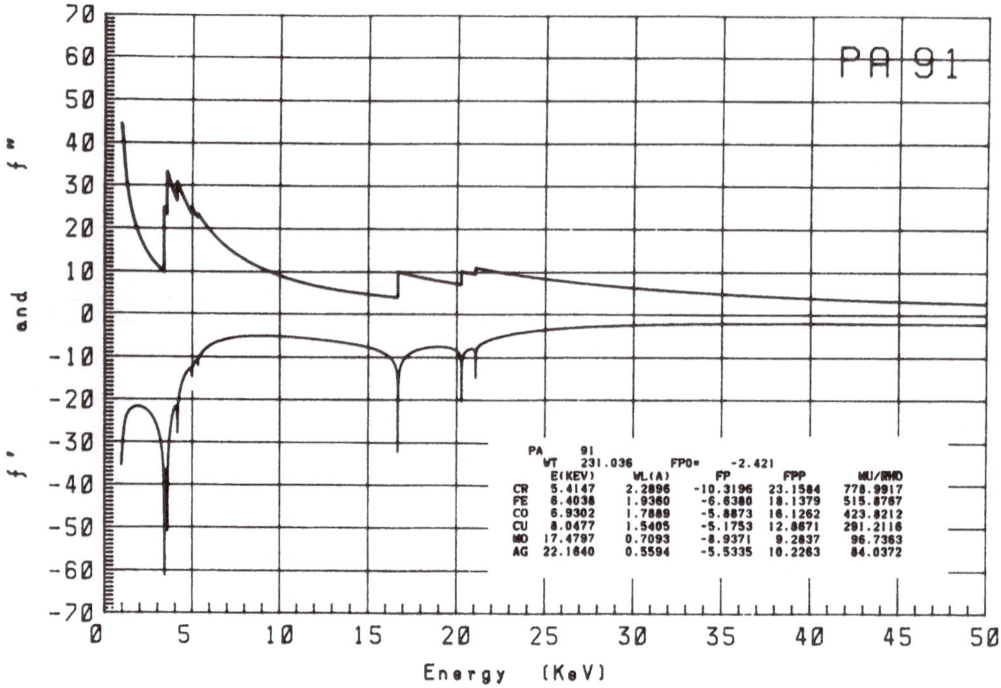

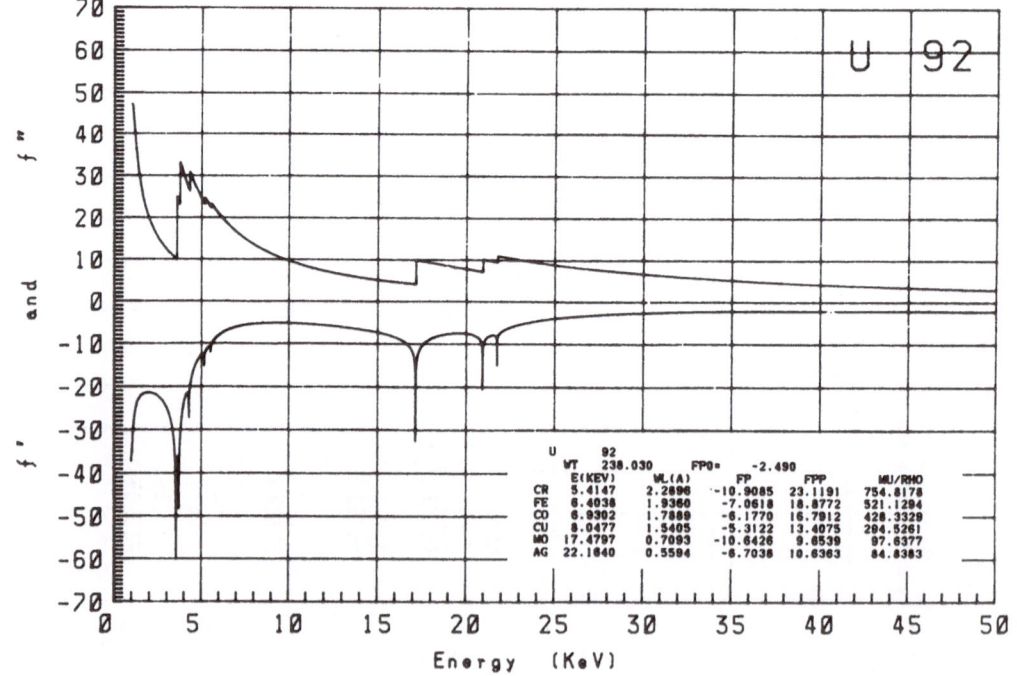

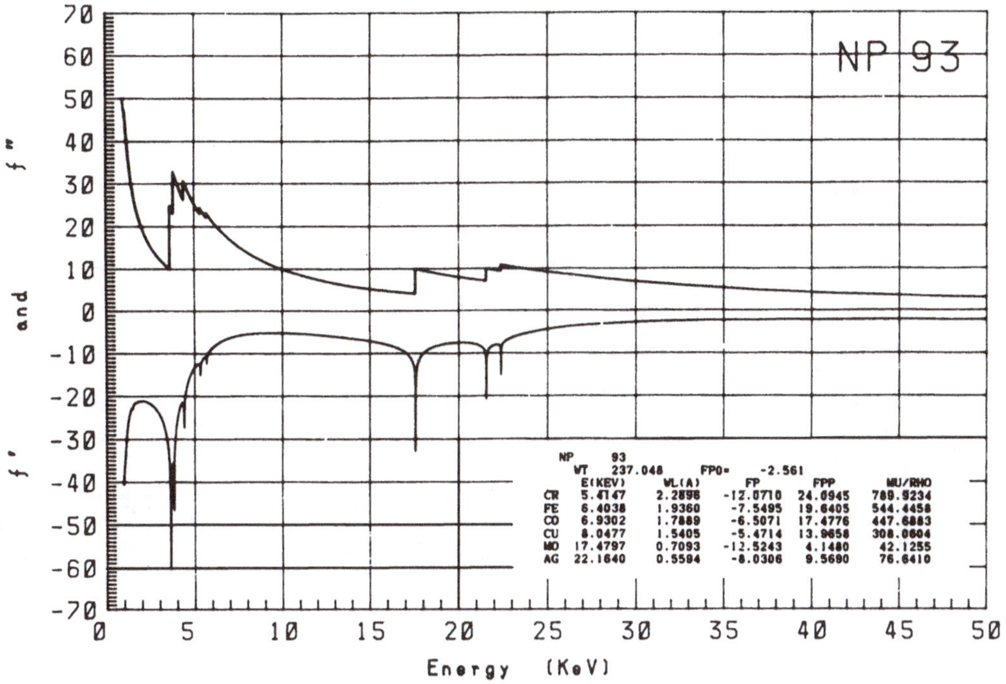

NP 93

NP	93			
WT	237.048	FP0=	-2.561	
	E(KEV)	WL(A)	FP	FPP	MU/RHO
CR	5.4147	2.2896	-12.0710	24.0945	789.9234
FE	6.4038	1.9360	-7.5495	19.6405	544.4458
CO	6.9302	1.7889	-6.5071	17.4776	447.6883
CU	8.0477	1.5405	-5.4714	13.9658	308.0604
MO	17.4797	0.7093	-12.5243	4.1480	42.1255
AG	22.1640	0.5594	-8.0306	9.5690	76.6410

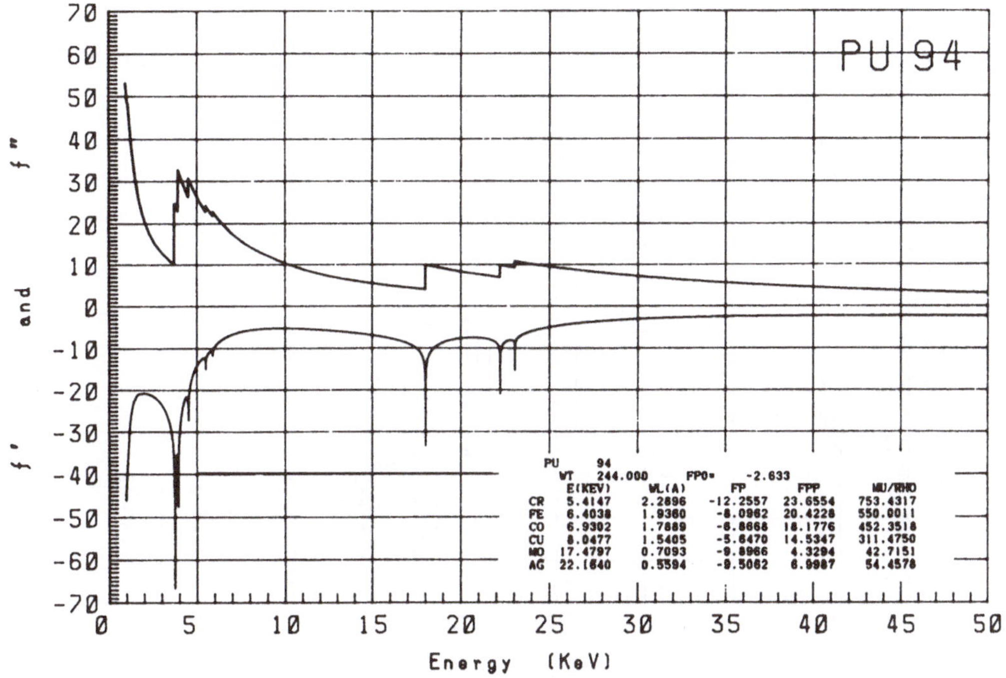

PU 94

PU	94			
WT	244.000	FP0=	-2.633	
	E(KEV)	WL(A)	FP	FPP	MU/RHO
CR	5.4147	2.2896	-12.2557	23.6554	753.4317
FE	6.4038	1.9360	-8.0962	20.4228	550.0011
CO	6.9302	1.7889	-6.8668	18.1776	452.3518
CU	8.0477	1.5405	-5.6470	14.5347	311.4750
MO	17.4797	0.7093	-8.8966	4.3294	42.7151
AG	22.1640	0.5594	-9.5082	6.9987	54.4578

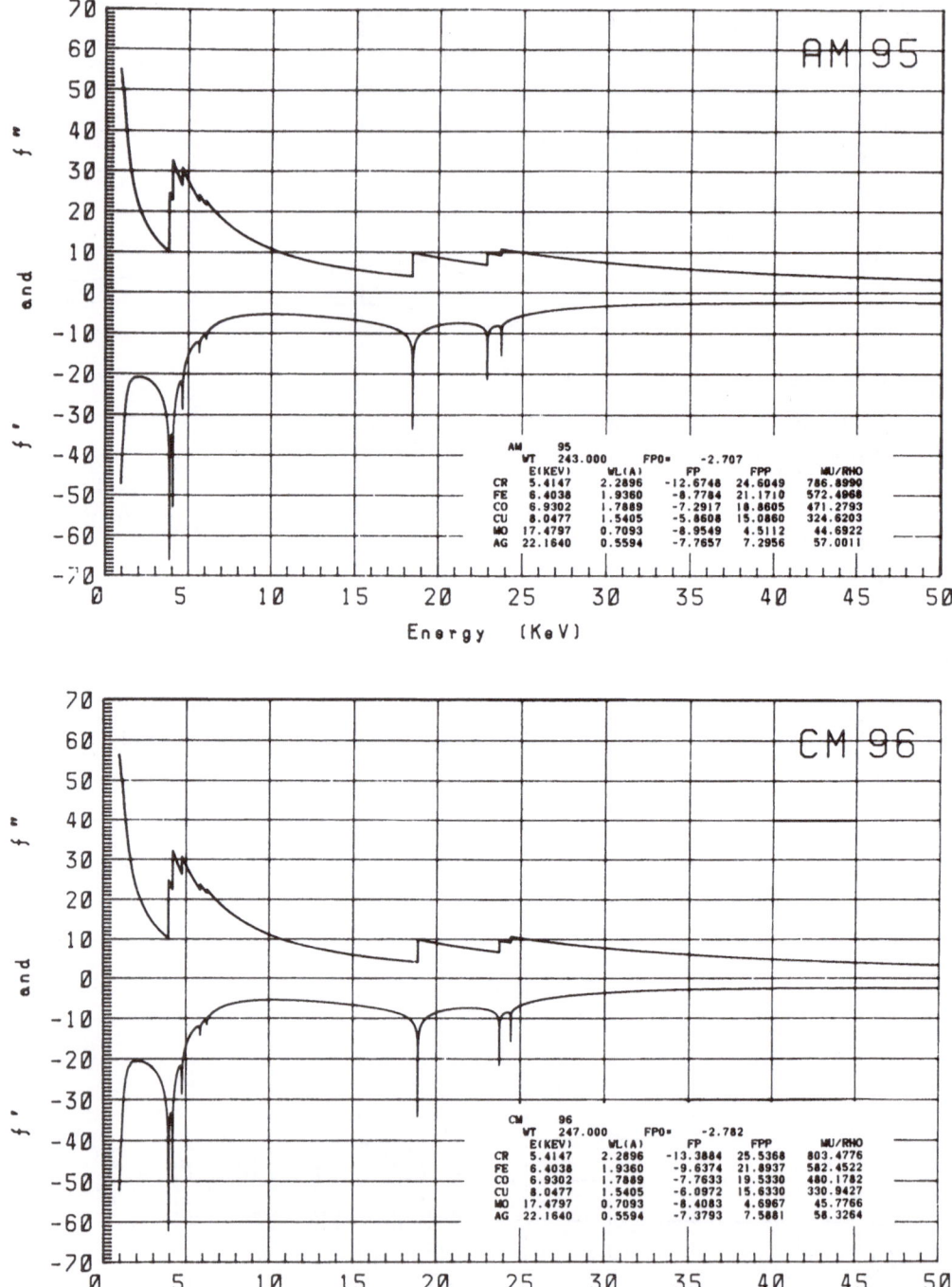

AM 95

AM	95				
WT	243.000		FPO=	-2.707	
	E(KEV)	WL(A)	FP	FPP	MU/RHO
CR	5.4147	2.2896	-12.6748	24.6049	786.8990
FE	6.4038	1.9360	-8.7784	21.1710	572.4968
CO	6.9302	1.7889	-7.2917	18.8605	471.2793
CU	8.0477	1.5405	-5.8608	15.0860	324.6203
MO	17.4797	0.7093	-8.9549	4.5112	44.6922
AG	22.1640	0.5594	-7.7657	7.2956	57.0011

CM 96

CM	96				
WT	247.000		FPO=	-2.782	
	E(KEV)	WL(A)	FP	FPP	MU/RHO
CR	5.4147	2.2896	-13.3884	25.5368	803.4776
FE	6.4038	1.9360	-9.6374	21.8937	582.4522
CO	6.9302	1.7889	-7.7633	19.5330	480.1782
CU	8.0477	1.5405	-6.0972	15.6330	330.9427
MO	17.4797	0.7093	-8.4083	4.6967	45.7766
AG	22.1640	0.5594	-7.3793	7.5881	58.3264

BK 97
70
60
50
40
30
20
10
0
-10
-20
-30
-40
-50
-60
-70
f' and f"
BK 97
WT 247.000 FP0= -2.858
E(KEV) WL(A) FP FPP MU/RHO
CR 5.4147 2.2896 -14.6739 26.7986 843.1762
FE 6.4038 1.9360 -10.4740 21.9396 583.6730
CO 6.9302 1.7889 -8.4706 20.3800 501.0010
CU 8.0477 1.5405 -6.4140 16.3156 345.3921
MO 17.4797 0.7093 -7.9687 4.9077 47.8329
AG 22.1640 0.5594 -7.4658 7.9301 60.9554
0 5 10 15 20 25 30 35 40 45 50
Energy (KeV)

CF 98
70
60
50
40
30
20
10
0
-10
-20
-30
-40
-50
-60
-70
f' and f"
CF 98
WT 251.000 FP0= -2.936
E(KEV) WL(A) FP FPP MU/RHO
CR 5.4147 2.2896 -16.1079 27.8958 863.7112
FE 6.4038 1.9360 -11.5238 22.7822 596.4304
CO 6.9302 1.7889 -9.2208 21.1199 510.9150
CU 8.0477 1.5405 -6.7156 16.9281 352.6482
MO 17.4797 0.7093 -7.6610 5.1071 48.9827
AG 22.1640 0.5594 -7.6215 8.2455 62.3697
0 5 10 15 20 25 30 35 40 45 50
Energy (KeV)

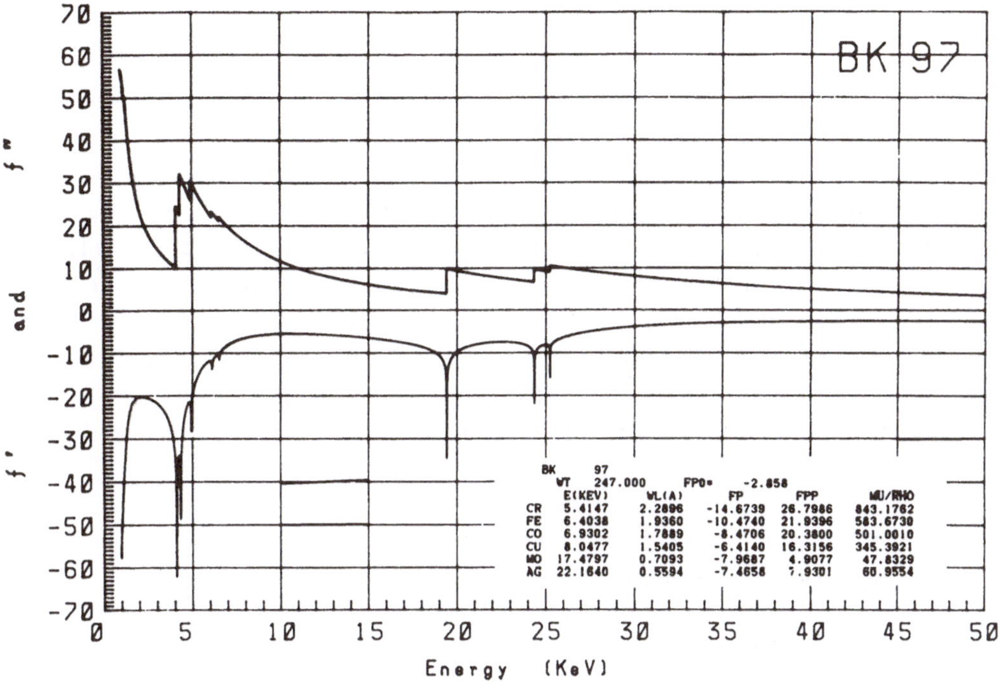

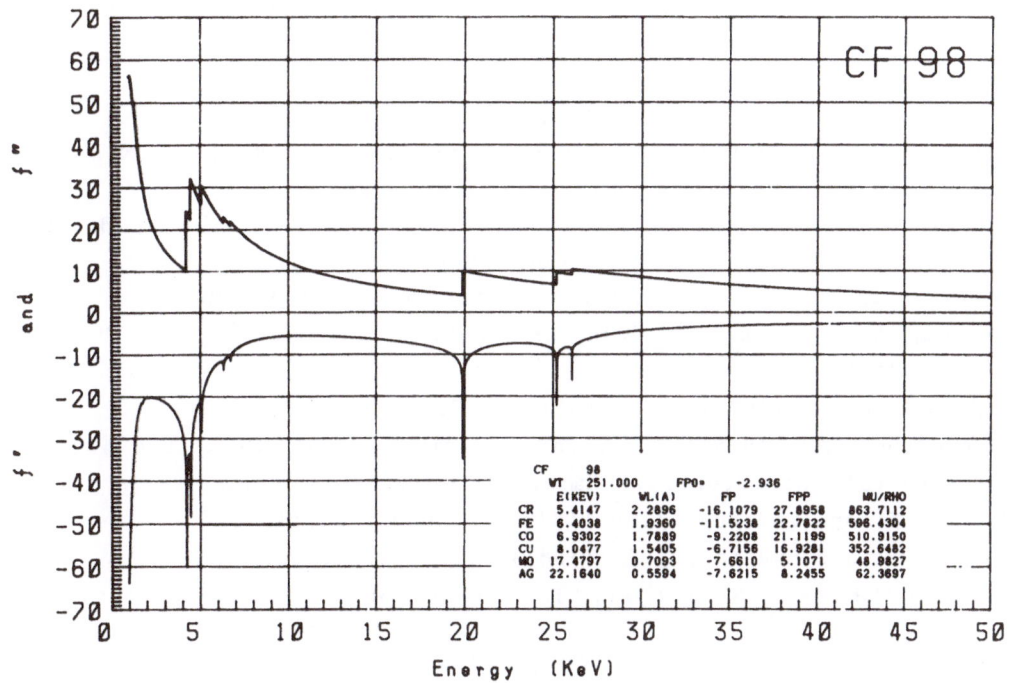

REFERENCES

Ashcroft, N.W. and Langreth, D.C. (1967), Phys. Rev. **159**,500.

Aur, S. and Egami, T. (1980), Proc. 4th Inter. Conf. on Liquid and Amorphous Metals, Grenoble, J. de Physique Suppl. **40**,C8-234.

Aur, S., Kofalt, D., Waseda, Y., Egami, T., Wang, R., Chen, H.S. and Teo, B.K. (1983a), Solid State Commun. **48**,111.

Aur, S., Kafalt, D., Waseda, Y., Egami, T., Chen, H.S., Teo, B.K. and Wang, R. (1983b), J. Nucl. Instrum. in press.

Avigon, M. and Falicov, L.M. (1974), J. Phys. F. **4**,1782.

Bacon, G.E. (1969), Acta cryst. **A25**,391.

Bacon, G.E. (1972), Acta Cryst. **A28**,357.

Baer, H.G. (1958), Z. Metallkde. **49**,614.

Beer S.Z. (Editor), (1972), Liquid Metals, Chemistry and Physics, Marcel-Dekker, New York.

Bellissent-Funel, M.C., Desre, P., Bellissent, R. and Tourand, G. (1977), J. Phys. F. **7**,2485.

Bertaut, F. (1950), Compt. Rend. **231**,88.

Bernal, J.D. (1959), Nature, **183**,141.

Betts, F., Bienenstock, A. and Ovshinsky, S.R. (1970), J. Non-Cryst. Solids, **6**,554.

Bhatia, A.B. (1977), Proc. 3rd Inter. Conf. on Liquid Metals, Bristol, The Inst. Phys. (London), Conf. Ser. No.30, p.21.

Bhatia, A.B. and Thornton, D.E. (1970), Phys. Rev. **B2**,3004.

Bhatia, A.B., Hargrove, W.H. and March, N.H. (1973), J. Phys. C. **6**,621.

Bienenstock, A. (1973), J. Non-Cryst. Solids, **11**,447.

Bienenstock, A. (1975), Proc. Inter. Symp. on the Structure of Non-Crystalline Materials, Cambridge Univ. Press, P.5.

Biggin, S. and Enderby, J.E. (1978), J. Phys. C. **11**,3577.

Biggin, S. and Enderby, J.E. (1981), J. Phys. C. **14**,703.

Blasie, J.K. and Stamatoff, J. (1981), Ann. Rev. Biophys. Bioeng. **10**,451.

Bletry, J. and Sadoc, J.F. (1975), J. Phys. F. **5**,L110.

Boiso, L., Texeira, J. and Stanley, H.E. (1981), Phys. Rev. Letters, **46**,597.

Bondot, P. (1974), Acta Cryst. **A30**,470.

Bonse, U. and Materlik, G. (1976), Z. Physik, **B24**,189.

Bonse, U., Hartmann-Lotsch, I., Lotsch, H. and Olthoff-Mienter, K. (1982), Z. Physik, **B47**,297.

Boyce, J.B. and Huberman, B.A. (1979), Phys. Rev. **51**,189.

Bradley, A.J. and Rodgers, J.W. (1934), Proc. Roy. Soc. **A144**,340.

Brysk, H. and Zerby, C.D. (1968), Phys. Rev. **171**,292.

Cargill III, G.S. (1975), Solid State Phys. edited by Ehrenreich, H., Seitz, F. and Turnbull, D. Academic Press, New York, **30**,227.

Cargill III, G.S. (1981), Proc. 4th Inter. Conf. on Rapidly Quenched Metals, Sendai, Japan Inst. Metals, Vol.1, p.389.

Cargill III, G.S. and Spaepen, F. (1981), J. Non-Cryst. Solids, **43**,91.

Cargill III, G.S. and Tsuei, C.C. (1978), Proc. 3rd Inter. Cong. on Rapidly Quenched Metals, The Metals Society (London), No.198, Vol.2, p.337.

Chen, H.S. (1980), Rep. Progr. Phys. **43**,353.

Chen, H.S. and Waseda, Y. (1979), phys. stat. sol. (a), **51**,593.

Chipman, D.R. (1955), J. Appl. Phys. **26**,1387.

Chipman, D.R., Jennings, L.D., and Giessen, B.C. (1978), Bull. Amer. Phys. Soc. **23**,467.

Cowley, J.M. (1950), J. Appl. Phys. **21**,24.

Cromer, D.T. (1965), Acta Cryst. **18**,17.

Cromer, D.T. and Liberman, D. (1970), J. Chem. Phys. **53**,1891.

Cromer, D.T. and Mann, J.B. (1967), J. Chem. Phys. **47**,1892.

Dauben, C.H. and Templeton, D.H. (1955), Acta Cryst. **8**,841.

DeRidder, R., DeSitter, J. and Amelinckx, S. (1974), phys. stat. sol. (a), **23**,615.

Derrin, J.Y. and Dupuy, J. (1976), Phys. Chem. Liquids, 5,71.

Edwards, F.G., Enderby, J.E., Howe, R.A., and Page, D.I. (1975), J. Phys. C. **8**,3483.

Edwards, F.G., Howe, R.A., Enderby, J.E. and Page, D.I. (1978), J. Phys. C. **11**,1053.

Egami, T. (1981a), Ann. New York Acad. Sci. **371**,238.

Egami, T. (1981b), Glassy Metals I edited by Guntherodt, H-J. and Beck, H., Springer-Verlag, Berlin, p.25.

Egami, T., Williams, R.S. and Waseda, Y. (1978), Proc. 3rd Inter. Conf. on Rapidly Quenched Metals, Brighton, The Metals Society (London), No.198, Vol.2, p.381.

Eisenberg, S., Jal, J.F., Dupuy, J., Chieux, P. and Knoll, W. (1982), Phil. Mag. **46**,195.

Eisenlohr, H. and Miller, L.J. (1954), Z. Physik, **136**,491 and 511.

Ellis, W.E. and Greiner, E.S. (1941), Trans. ASM, **29**,415.

Enderby, J.E., (1968), Physics of Simple Liquids edited by Temperley, N.H., Rowlinson, J.S. and Rushbrooke, G.S., North-Holland, Amsterdam, p.612.

Enderby, J.E., Egelstaff, P.A. and North, D.M. (1966), Phil. Mag. **14**,961.

Enderby, J.E. and Neilson, G.W. (1981), Rep. Progr. Phys. **44**,593.

Faber, T.E. (1972), An Introduction to the Theory of Liquid Metals, Cambridge Univ. Press.

Faber, T.E. and Ziman, J.M. (1965), Phil. Mag. **11**,153.

Finbak, C. (1949), Acta Chem. Scand. **3**, 1279 and 1293.

Fukamachi, T. (1977), J. Cryst. Soc. Japan, **19**,51.

Fukamachi, T., Hosoya, S., Kawamura, T., Hunter, S. Nakama, Y. (1978), Japanese J. Appl. Phys. Suppl. **17-2**,326.

Fukamachi,T., Hosoya, S., Kawamura, T. and Okunuki, M. (1979), Acta Cryst. **A35**,104.

Fuoss, P.H., Eisenberger, P., Warburton, W.K. and Bienenstock, A. (1981), Phys. Rev. Lett. **46**,1537.

Fuoss, P.H., Warburton, W.K. and Bienenstock, A. (1980), Proc. 8th Inter. Conf. on Amorphous and Liquid Semiconductors, Cambridge, J. Non-Cryst. **35/36**,1233.

Funke, K. (1976), Progr. Solid State Chem. **11**,365.

Furukawa, K. (1962), Rep. Progr. Phys. **25**,395.

Gerward, L., Thuesen, G., Jensen, M.S. and Alstrup, I. (1979), Acta Cryst. **A35**,852.

Gopolakrishnan, P.S. and Ramaseshan, S. (1975), Acta Cryst. **A31**,S159.

Graham Jr, C.D. and Egami, T. (1978), Ann. Rev. Mater. Sci. **8**,423.

Groubert, E. and Regis, R. (1967), Ann. de Physique, **2**,305.

Guntherodt, H-J. and Beck, H. (Editors), (1981), Glassy Metals I, Springer-Verlag, Berlin.

Hagenmuller, P.and van Gool,W. (Editors),(1981), Solid Electrolytes, Academic Press, New York.

Halder, N.C. and Wagner, C.N.J. (1967), J. Chem. Phys. 47,4385.

Hayes, T.M. and Boyce, J.B., (1982), Solid State Phys. edited by Ehrenreich, H., Seitz, F. and Turnbull, D. Academic Press, New York, 37,173.

Hayes, T.M., Boyce, J.B. and Beeby, J.L. (1978), J. Phys. C. 11,2931.

Hill, T.L. (1956), Statistical Mechanics, McGraw-Hill, New York.

Honl, H. (1933), Z. Phys. 84,1 and Ann. Physik, 18,625.

Hoshino, S. (1957), J. Phys. Soc. Japan, 12,315.

Hoshino, S. (1978), Solid State Phys. (Japan), 12,315.

Hoshino, S., Sakuma, T. and Fujii, Y. (1977), Solid State Commun. 22,763.

Hosoya, S. (1970), Bull. Phys. Soc. Japan, 25,110 and 288.

Hosoya, S. (1977), J. Cryst. Soc. Japan, 19,68.

Hosoya, S. (1979), Rigaku-Denki-Journal, 10,7.

Hosoya, S. and Fukamachi, T. (1973), J. Appl. Cryst. 6,396.

Hosoya, S., Kawamura, T. and Fukamachi, T. (1978), Bull. Japanese, Appl. Phys. 47,708.

Hosoya, S. and Yamagishi, T. (1966), J. Phys. Soc. Japan, 21,2638,

Huijben, M.J., Lee, T., Reimart, W. and van der Lugt, W. (1977), J. Phys. F. 7,L119.

International Tables for X-ray Crystallographys (1974), Vol. IV edited by Ibers, J.A. and Hamilton, W.C. The Kynoch Press, Birmingham.

James, R.W. (1954), The Optical Principles of the Diffraction of X-rays, G. Bell & Sons, London.

Kawamura, T. and Fukamachi, T. (1978), Japanese J. Appl. Phys. Suppl. 17-2,228.

Keating, D.T. (1963), J. Appl. Phys. 34,923.

Kortright, J. and Bienenstock, A. (1984), Proc. 5th Inter. Conf. on Liquid and Amorphous Metals, Los Angeles, J. Non-Cryst. Solids, 61/62,273.

Krogh-Moe, J. (1956), Acta Cryst. 9,951.

Krogh-Moe, J. (1966), Acta Chem. Scand. 20,2890.

Klug, H.P. and Alexander, L.E. (1974), X-ray Diffraction Procedures for Polycrystalline and Amorphous Materials, 2nd Edition, Wiley, New York.

Kudielka, H. and Moller, H. (1963), Z. Metallkude. **118**,S213.

Lamparter, P., Sperl, W., Steeb, S. and Bletry, J. (1982), Z. Naturforsch. **37a**,1223.

Landau, L.D. and Lifshitz, E.M. (1968), Statistical Physics, Pergamon Press, Oxford.

Lee, P.A., and Teo, B.K. (Editors), (1981), EXAFS Spectroscopy, Plenum Press, New York.

Lee, P.A. Citrin, P.H., Eisenberger, P. amd Kincaid, B.M. (1981), Rev. Mod. Phys. **53**,761.

Mahan, G.D. and Roth, W.L. (Editors), (1976), Superionic Conductors, Plenum Press, New York.

Malet, G., Cabos, C., Escande, A. and Delard, P. (1973), J. Appl. Cryst. **6**,139.

Mariano, A.N. and Hanneman, R.E. (1963), J. Appl. Phys. **34**,384.

Masumoto, T., Fukunaga, T. and Suzuki, K, (1978),Bull. Amer. Phys. Soc. **23**,467; (1980), Sci. Rep. Res. Inst. Tohoku University, **28A**,208.

McGreeby, R.L. and Mitchell, E.W.J. (1982), J. Phys. C. **15**,5537.

Meisel, L.V. and Cote, P.J. (1977), Phys. Rev. **B15**,2970.

Mitchell, E.W.J., Ponet, P.F.J. and Stewart, R.J. (1976), Phil. Mag. **34**,721.

Miyake, S. (1969), X-ray Diffraction, Asakura Book Co. Ltd., Tokyo.

Mizoguchi, T., Kudo, T., Irisawa, T., Watanabe, N., Niimura, N., Misawa, M. and Suzuki, K., (1978), Proc. 3rd. Inter. Conf. on Rapidly Quenched Metals, Brighton, The Metals Society (London), No.198, Vol. 2, p.384; and J. Phys. Soc. Japan, **45**(1978),1773.

Munro, R.G. (1982), Phys. Rev. **B25**,5037.

Nishikawa, S. and Matsukawa, K. (1928), Proc. Imp. Acad. Japan, **4**,96.

Nold, E., Lamparter, P., Olbrich, H., Rainer-Harbach, G. and Steeb, S. (1981), Z. Naturforsch. **36a**,1032.

Norman, N. (1957), Acta Cryst. **10**,370.

Okazaki, H. (1967), J. Phys. Soc. Japan, 23,355.

O'Leary, W.P. (1975), J. Phys. F. 5,L175.

Oyanagi, H. and Hosoya, S. (1980), J. Cryst. Soc. Japan, 22,57.

Paasche, F., Olbrich, H., Rainer-Harbach, G., Lamparter, P. and Steeb, S. (1982), Z. Naturforsch. 37a,1215.

Page, D.I. and Mika, K. (1971), J. Phys. C. 4,3034.

Parratt, L.G. and Hempstead, C.F. (1954), Phys. Rev. 94,1593.

Peterson, S.W. and Smith, H.G. (1961), Phys. Rev. Lett. 6,7.

Rahman, A. (1965), J. Chem. Phys. 42,3540.

Ramaseshan, S. (1964), Advanced Methods of Crystallography edited by Ramachandran, G.N., Academic Press, London, p.64.

Ramaseshan, S. and Abraham, S.C. (Editors), (1975), Anomalous Scattering, (Inter. Union of Crystallography), Munksgaard, Copenhagen.

Ramesh, T.G. and Ramaseshan, S. (1971), J. Phys. C. 4,3029.

Richardson, F.D. (1974), Physical Chemistry of Melts in Metallurgy, Academic Press, London.

Ruppersberg, H. and Egger, H. (1975), J. Chem. Phys. 63,4095.

Sadoc, J.F. and Dixmier, J. (1976), Proc. 2nd Inter. Conf. on Rapidly Quenched Metals, Cambridge, Mass. Mater. Sci. Eng. 23,187.

Sakuma, T., Iida, T., Honma, K. and Okazaki, H. (1977), J. Phys. Soc. Japan, 43,538.

Sandstrom, D.R. (1979), J. Chem. Phys. 71,2381.

Sayers, D.E., Stern, E.A. and Lytle, F.W. (1971), Phys. Rev. Lett. 27,1204.

Schlapbach, L. (1974), Phys. Kond. Matter, 19,189.

Shevchik, N.J. (1977), Phil. Mag. 35,805 and 1289.

Sinclair, R.N., Jonson, D.A.G., Dore, J.C., Clark, J.H. and Wright, A.C. (1974), Nucl. Intrum. Meth. 117,445.

Sirota, N.N. (1969), Acta Cryst. A25,223.

Skolnick, L.P., Kudo, S. and Lavine, L.R. (1958), J. Appl. Phys. 29,198.

Soper, A.K., Neilson, G.W., Enderby, J.E. and Howe, R.A. (1977), J. Phys. C. 10,1793.

Srinivasan, R.(1972),Adv.in Struc. Rev. by Diffrac. Method, 4,105.

Stern, E.A., Sayers, D.E. and Lytle, F.W. (1975), Phys. Rev. 11,4836.

Strock, J.W. (1934), Z. Phys. Chem. B25,441.

Sugawara, T. (1951), Sci. Rep. Res. Inst. Tohoku University, Sendai, 3A,39.

Suzuki, K. (1976), Bunsen-Gesll. Phys. Chem. 80,689.

Suzuki, K., Fukunaga, T., Misawa, M. and Masumoto, T. (1976), Sci. Rep. Res. Inst. Tohoku University, 26A,1. and Marer. Sci. Eng. (1976),23,215.

Svab, E., Forgacs, F., Hajdu, F., Kroo, N. and Takacs, J. (1978), J. Non-Cryst. Solids, 45,1773.

Takeda, S., Tamaki, S. and Waseda, Y. (1983), J. Phys. Soc. Japan, 52,2062.

Tamaki, S. and Waseda Y. (1980), J. Cryst. Soc. Japan, 22,20.

Tamaki, S., Tsuchiya, Y., Cusack, N.E., Waseda, Y. and Jacob, K.T. (1989), J. Phys. F. 10,L109.

Temperley, N.H., Rowlinson, J.S. and Rushbrooke, G.S. (Editors), (1968), Physics of Simple Liquids, North-Holland, Amsterdam.

Templeton, D.H., Templeton, L.K., Phillips, J.C. and Hodgson, K.O. (1980), Acta Cryst. A36,436.

Teo, B.K., Chen, H.S., Wang, R. and Antonio, M.R. (1983), J. Non-Cryst. Solids, 58,249.

Tsuchiya, Y. Tamaki, S., Waseda, Y. and Toguri, J.M. (1978), J. Phys. C. 11,651.

Tsuchiya, Y., Tamaki, S., Waseda, Y., (1979), J. Phys. C. 12,5361 and L93.

Vashishta, P. and Rahman, A. (1978), Phys. Rev. Lett. 40,1837.

Vineyard, G.H. (1958), Liquid Metals and Solidification, American Society for Metals, Cleveland, p.1.

Wagenfeld, H. (1966), Phys. Rev., 144,216.

Wagenfeld, H. (1975), Anomalous Scattering edited by Ramaseshan, S. and Abraham, S.C. (Inter, Union of Crystallography), Munksgaard, Copenhagen, p.13.

Wagner, C.N.J. (1972), Liquid Metals, Chemistry and Physics edited by Beer, S.Z., Marcel-Dekker, New York, p.258.

Wagner, C.N.J. (1978), J. Non-Cryst. Solids, 31,1.

Wakelin, R.J. and Yates, E.L. (1953), Proc. Phys. Soc. **B66**,221.

Waseda, Y. (1980), The Structure of Non-Crystalline Materials, McGraw-Hill, New York.

Waseda, Y. (1981), Progr. Mater. Sci. 25,1.

Waseda, Y. and Hirata, K. (1975), Bull. Res. Inst. Min. Met. Tohoku University, **31**,9.

Waseda, Y. and Tamaki, S. (1975), Phil. Mag. 32,951.

Waseda, Y. and Tamaki, S. (1976), Z. Physik, **B23**,315.

Waseda, Y. and Tamaki, S. (1977), J. Phys. F. 7,L151.

Waseda, Y. and Toguri, J.M. (1978), Proc. Inter. Conf. on Phys. of Transition Metals, Toronto, The Inst. Phys. (London), Conf. Ser. No.39, p.432.

Waseda, Y. and Sakuma, T. (1982), Bull. Japan Inst. Metals, **21**,69 and Metal Phys. Seminar (Japan), **5**,201.

Waseda, Y., Masumoto, T. and Tamaki, S. (1977), Proc. 3rd Inter. Conf. on Liquid Metals, Bristol, The Inst. Phys. (London), Conf. Ser. No.30, p.268.

Waseda, Y., Ohta, H., Aur, S. and Egami, T. (1984), Z. Naturforsch. **39a**, in press and SCM Report-84-001, Res. Inst. Min. Met. Tohoku University.

Warren, B.E. (1969), X-ray Diffraction, Addison-Wesley, Reading, Mass.

Warren, B.E., Averbach, B.L. and Roberts, B.W. (1951), J. Appl. Phys. **22**,1493.

Weiss, R.J. (1966), X-ray Determination of Electron Distributions, North-Holland, Amsterdam.

Wong, J.(1980), Glassy Metals I edited by Guntherodt,H-J.and Beck, H., Springer-Verlag, Berlin, p.45.

Wright, A.C. and Leadbetter, A.J. (1976), Phys. Chem. Glasses, 17,122.

Yamada, K., Endoh, Y. Ishikawa, Y. and Watanabe, N. (1980), J. Phys. Soc. Japan, **48**,922.

Yazawa, A., Takeda, Y. and Waseda, Y. (1981), Can. Met. Quart. **20**,129. and Report of the 28th Congreth of IUPAC, Vancouver.

Zachariasen, W.H. (1965), Acta Cryst. **18**,714.

Ziman, J.M. (1979), Models for Disorder, Cambridge Univ. Press.

Subject Index

Lecture Notes in Physics

Light Scattering in Solids I

Introductory Concepts

Editor: **M. Cardona**
With contributions by numerous experts

2nd corrected and updated edition. 1983. 111 figures.
XV, 363 pages
(Topics in Applied Physics, Volume 8)
ISBN 3-540-11913-2

Amorphous Semiconductors

Editor: **M. H. Brodsky**
With contributions by numerous experts
1979. 181 figures, 5 tables. XVI, 337 pages
(Topics in Applied Physics, Volume 36)
ISBN 3-540-09496-2

Glassy Metals I

Ionic Structure, Electronic Transport, and Crystallization

Editors: **H.-J. Güntherodt, H. Beck**
With contributions by numerous experts

1981. 119 figures. XIV, 267 pages
(Topics in Applied Physics, Volume 46)
ISBN 3-540-10440-2

Physics of Superionic Conductors

Editor: **M. B. Salamon**
With contributions by numerous experts

1979. 101 figures, 13 tables. XII, 255 pages
(Topics in Current Physics, Volume 15)
ISBN 3-540-09333-8

Dissipative Systems in Quantum Optics

**Resonance Fluorescence, Optical Bistability,
Superfluorescence**

Editor: **R. Bonifacio**
With contributions by numerous experts

1982. 60 figures. XI, 151 pages
ISBN 3-540-11062-3

Springer-Verlag
Berlin
Heidelberg
New York
Tokyo

Selected Issues from
Lecture Notes in Mathematics